D0055190

THE TURBULENT UNIVERSE

PAUL KURTZ

THE TURBULENT UNIVERSE

59 John Glenn Drive
Amherst, New York 14228–2119

Published 2013 by Prometheus Books

The Turbulent Universe. Copyright © 2013 by the Estate of Paul Kurtz. All rights reserved. No part of this publication may be reproduced, stored in a retrieval system, or transmitted in any form or by any means, digital, electronic, mechanical, photocopying, recording, or otherwise, or conveyed via the Internet or a website without prior written permission of the publisher, except in the case of brief quotations embodied in critical articles and reviews.

Trademarks: In an effort to acknowledge trademarked names of products mentioned in this work, we have placed ® or ™ after the product name in the first instance of its use in each chapter. Subsequent mentions of the name within a given chapter appear without the symbol.

Cover image © 2013 Photodisc, Inc.
Cover design by Grace M. Conti-Zilsberger

Inquiries should be addressed to
Prometheus Books
59 John Glenn Drive
Amherst, New York 14228–2119
VOICE: 716–691–0133 • FAX: 716–691–0137
WWW.PROMETHEUSBOOKS.COM

17 16 15 14 13 5 4 3 2 1

Library of Congress Cataloging-in-Publication Data Pending

Kurtz, Paul, 1925-2012.
The turbulent universe / by Paul Kurtz.
p. cm.
Includes bibliographical references.
ISBN 978-1-61614-735-8 (cloth : alk. paper)
ISBN 978-1-61614-736-5 (ebook)
1. Secular humanism. I. Title.

B821.K87 2013
144—dc23

2012048877

Printed in the United States of America on acid-free paper

CONTENTS

OVERTURE

PONDERING THE NIGHT SKY

I begin this book sitting in my study in a chateau overlooking a small French village below. Looking out of the window I can see on a nearby mountain the city of Grasse, the former perfume center in Southern France. As the Sun begins to set, the streetlamps and house lights in Grasse begin to flicker. In the distance is Cannes, the film capital of Europe, bordering on the still-visible blue Mediterranean Sea. The chateau was built by my wife's father several decades ago. It is perched on a small mountain in the Alpes-Maritime. Above us is Castellaras, a retreat for retirees. Below is a still-visible color-splashed valley. As the dusk settles in, I can still see an olive orchard right below me, palm and pine trees, and exotic flowers and plants. There are few lamps lit at the moment on the property surrounding the chateau, and so as the twilight descends, I am able to view the glorious night sky.

On a clear night the heavens offer a fabulous display of galaxies, stars, planets, and an occasional meteor shower. I can remember seeing the Hale-Bopp comet in 1997 with its spectacular sparkling tail. Below is the village of Mouans-Sartoux, where my wife's ancestors lived for over five centuries, ever since 1495, when they were invited by the Seigneur of Grasse to settle after the Black Death in the eleventh century had devastated and largely depopulated the area. They came from Zaragoza, Spain, at first—victims of the Inquisition—and they found refuge in Genoa, Italy. They accepted the invitation gladly. My father-in-law showed me the earliest gravestone markers in the cemetery, barely legible, with the family names Vial (there was a Claude Vial in the first group of invitees) and Vidal prominent. There were so many intermarriages since then that the bloodlines have been intermixed with the stocks of many ethnicities.

I have been coming to the Côte d'Azur (the French Riviera) for every summer and on special holidays for half a century with my wife and our children and grandchildren. In the early years, I remember the stunning growth in the spring of jasmine and lavender, whose delicate fragrances saturated the air. Farmers harvested the flowers in those days and extracted their essences for the lovely perfume that was bottled and sold to fashionable ladies in Paris, Toronto, and Los Angeles. We would often visit the resort city of Cannes with its gleaming white hotels and fashionable boutiques, cafes, and restaurants, where we would on occasion dine on fresh caught seafood (*poisson de mer*) and wines of the province.

Mouans and Sartoux were separate villages that merged in the mid-nineteenth century. The original homes at the center of town were built of tan-hued solid stones mined from a nearby quarry. The town's narrow streets are still cobbled and gutted by generations of horseshoes and cartwheels; Mouans-Sartoux has since been modernized and spiffed up a bit with some fancy shops for the increasing number of tourists.

I walk to town almost daily to bring back fresh-baked loaves of still-hot bread and to purchase *Le Monde* or *Le Figaro* and the *International Herald Tribune*. The avenue that I take is called Route Napoléon, the road that Napoleon took (from Cannes to Grasse to Grenoble) in March 1815 to raise an army in France a second time after returning from his exile at Elba, a small island in the Mediterranean. And so I speculate about viewing things from a historical perspective and what this portends for understanding not only seasoned cultures such as France but also the multiplicity of human civilizations of the past, the evolving biosphere, and the vastness of the universe at large.

Goethe observed that one must speak two languages to fully understand one. Who can express phrases or expletives such as *voilà* or *oh la la* or *la vie en rose* in English? How can we translate the German words *achtung* or *ja wohl* into English or French? The English expressions *my word!* or *lovely* cannot be easily duplicated elsewhere. Idioms are idioms. Does *merde* have an exact equivalent in other languages? I doubt it.

Similarly, one has to live in at least two cultures to get some perspective on one's own idiosyncratic cultural foibles. That is why being married to a French woman and visiting France frequently and living there from time to time—seeing my mother-in-law every year when she was alive—gives me a new understanding (*verstehen*) of the incomprehensibility of her love for flowers or gourmet cuisine, the sauces and truffles. That is why my wife is torn between America, her adopted land where she has lived two-thirds of her life, and France, her homeland, still deep within her heart.

Another contrast is that the French hug and kiss family members, friends, and acquaintances whenever they arrive home, meet them, or say *au revoir*. This is also a mark of civility on diplomatic occasions between heads of state and foreign diplomats. The first time I met my father-in-law, my wife whispered into my ear, "Please embrace and kiss my father" (who had been a colonel in the French army and then a minister of finance and an entrepreneur in French Equatorial Africa). "What?" I said at first, shocked at what she asked me to do. But I said, "*mais oui*," and obliged, though my kiss was in the air, not on his cheeks. Americans may embrace women, but they generally only shake hands with men. But I now prefer the French custom as a symbol of politeness, respect, or affection.

Back to the heavenly bodies. As I gaze into the night sky, I think about how prehistoric humans and the countless generations who followed them through the millennia were no doubt impressed by the mysterious sky. They may have wondered about the sublime spectacle that they encountered. Surely they were astonished by the brightly lit panoply above, whether peering from the mountains, as I now am, or looking up while sleeping on the open pampas or desert sands, nestled at the top of the canopy of trees in the jungles, sheltered in the mouths of caves, or viewing the sky from the frozen tundra of the North. They were comforted by the Sun by day, the bearer of warmth and light, and by the phases of the Moon by night. The bright planets they saw (only much later identified as such), the vast number of flickering stars, the occasional meteor showers or rare comet or eclipse no doubt aroused their imagination. They were frightened by bolts of lightning and the pounding of thunder, the torrents of rain or blinding snow. And they perhaps pondered what it all meant, if anything. The staggering beauty of nature was a source of mystery and wonder.

Their lives were often threatened by dangers. They were often wracked by hunger or thirst, or frightened by ferocious beasts or marauding tribes. When they were wounded, fell sick, or faced the sudden death of a loved one, they may have cried out, "Is there some place in this vast scheme of things for me and my loved ones?" Many no doubt longed for some answers to the puzzle of existence. The shamans and magicians began to weave stories handed down by oral tradition from their forebears. Beliefs developed that there was a hidden realm beyond this world, controlled by unseen deities, and that we humans must propitiate them if we are to be shielded from the torments of brutish existence. The earliest inhabitants of France undoubtedly were also perplexed or frightened by the world of wild beasts that they encountered almost daily. Life was hazardous, but they survived.

A graphic illustration of how prehistoric people lived is found in the many caves discovered in Europe. The Chauvet-Pont-d'arc cave, about 150 kilometers northeast of Mouans-Sartoux, was discovered in 1994 by Jean Marie Chauvet and his two friends, who were fascinated by speleology (cave exploration). The cave is now named after Chauvet.

Chauvet and his colleagues were walking through fairly well-traversed hills in the Ardèche Valley region when they chanced upon a draft of cold air coming from a small cavity in a cliff behind some fallen rocks. They dug a passage, managed to crawl into it, and found themselves near a hidden shaft. Using a rope ladder, they eventually dropped down to a vast cave vault. They were overwhelmed when they entered a huge chamber, where they encountered some four hundred stunning drawings—some etched on the walls of the cave, some in black, and others in red ochre or tan—of handsome, well-drawn horses, bison, reindeer, rhinos, hyenas, and other animals. There were also human palm prints on the rock walls of the caves, perhaps as signatures. Animal bones were strewn about the floor of the cave; the remains of beasts probably eaten by the inhabitants, although there were also many skeletons of extinct giant cave bears. There were other large chambers adjacent. Carbon dating of the charcoal on the walls indicated that the drawings were etched approximately thirty-two thousand years ago. It is estimated that the artifacts are between thirty-five and thirty-six thousand years old, which suggests different episodes of occupancy during the Paleolithic period. At least one of these glittering drawings is of a female with exposed breasts and vulva; some etchings were part human and part animal. Obviously prehistoric humans needed to hunt for food, and they also encountered wild animals that attempted to devour them.

Some anthropologists have interpreted the drawings as reflecting a kind of hunter-magic significance, perhaps religious shamanism. The drawings may be a kind of propitiation for successful future hunts. The art strongly suggests that the people who drew them were cognitively sophisticated human beings, not fierce primitive cave folk with clubs in hand, but capable of creative drawings of artistic merit.

Perhaps in drawing the likeness of an animal, they believed that they might increase the hunter's catch. There were paintings of humans, partly transformed into animals, with the head and horns of a bison; one can only speculate about their meaning. Yet the ability of these primitive artists is impressive—it marks in a very real sense the first stirrings of culture. There are different sites in various caves in France and Spain, with artifacts created thousands of years apart. The discovery of tools and weapons used by prehistoric people is of considerable significance, because one of the distinctive characteristics of humans is their ability to invent tools and pass them on to future generations. These instruments mark a cultural advance, as they provide new ways for coping with nature and improving the conditions of life. The caves no doubt provided refuge, especially during the Ice Age glaciers, but they were dangerous places, also used by bears, lions, mammoths, and other animals.

The universe is mysterious to humans, a source of wonder and imagination, whether it is the heavenly skies above or the world below. Early humans attempted to fathom as best they could what they encountered in an effort to make sense of it.

They no doubt worried about illness and death. Various early cultures appeared to hope for eternal life and immortality of the soul, such as the Egyptian pharaohs who—much later—built huge pyramids to ensure their preservation in an afterlife. Other cultures spun parables of salvation and redemption by superhuman gods who sent emissaries promising rescue from death. As the rudiments of early religions developed, priests practiced incantations demanding sacrifice and obedience to their commandments. Out of these primitive yearnings prephilosophical and prescientific cosmologies of divine powers and gods were created, wishful reveries of a mysterious world beyond this life. These theologies postulated omnipotent deities who were responsible for the universe and provided its structure and order. Priests and poets practiced rituals, promising to heal those who suffered and save those who died. Religion thus grew out of the brute existential encounters with adversity and offered balm for the grieving heart.

Traditional theism provided a supernatural order to the universe. What is, was, or will be is due to a divine act of creation. God was the cause of nature. Each part had a purpose and fulfilled a design. The duty of humans in the scheme of things was to adore God and obey his commandments. The relationship of humans to God was the central theme of history, and if men and women were to achieve eternal rescue from the vicissitudes of fortune, then faith in an all-powerful deity had to dominate all other human concerns.

Today modern men and women need not begin their quest for the meaning of existence by contemplating the gods; they are not overwhelmed by faith. Nor should the rejection of God be the necessary starting point of inquiry—as many militant atheists assume. We should not begin our inquiry with critiques of the God hypothesis; rather *we should start afresh with nature itself, as we encounter it*. We need to enter into the world of nature and try to understand it without any of the presuppositions of earlier human civilizations. Our starting point should not be religion but science—there is a successful modern saga of this kind of inquiry.

In the light of recent astronomical discoveries, we are forced to abandon the primitive view that the sky is placid and the constellations unchanging. It is anything but that. We have viewed the deaths of stars from supernova explosions, which lead to the births of myriad new stars enriched by earlier stellar generations. These were seemingly hatched out of cocoons of stardust and condensed out of murky clouds of helium and hydrogen. The evolution of life on the planet Earth and the development of cultural history is the key to understanding who and what we are, and our prospects for the future.

NEO-HUMANISM

One may pose the question: Will the contemporary age of science ultimately spell the eclipse of theism and the "twilight of the gods"? It is difficult to predict the future course of human civilization. Can science bring about the dawn of a new era in which humans fully realize that they exist in a universe without God, and that they need to rely on their own resources of creative intelligence and courage if they are to survive and thrive? Every generation will undoubtedly be confronted with difficult problems and challenges. The danger is that this may lead to a new "failure of nerve" and the reappearance of the ancient myths of consolation in order to assuage human fear and foreboding.

A new movement called the "New Atheism" has suddenly sprouted up. It rejects belief in God as contrary to what the sciences tell us about the universe.[1] Will this forceful denunciation of the gods prevail? It no doubt provides a necessary antidote to ancient theological mythologies. Given the long-standing propensity for the gods to persist in human history, it will be a tortuous process to overcome them. In my view, the New Atheism (or agnosticism or skepticism) can help bring about the flowering of a self-reliant humanism in all of its grandeur. But only if, I submit, it is accompanied by a commitment to new ethical ideals and values that provide a genuine alternative to religion. This presupposes that secularism, which has been rapidly growing in Europe, the United States, and Asia, will continue to spread, and that this will contribute to the growth of a *new and vibrant secular humanism*. But it must adopt a genuine new approach. This I call "Neo-Humanism," which, although it is committed to secular humanism and scientific rationalism, is receptive to other points of view and is willing to work with others interested in developing a better world. It is not militantly antireligious. Although it is critical of the claims of theistic religion, its critiques are dignified and rational. It is skeptical of such claims, but it aims to be inclusive by appealing to a wider range of people.

If skeptical doubts about the ancient monotheistic religions eventually penetrate social awareness, then the otherworldly religions will begin to lose their influence on society. But if this is to prevail, then Neo-Humanists need to concentrate on improving the things of this world rather than simply combating the illusions of supernaturalism. Accordingly, the central question is whether a new form of planetary humanism can be developed in which humans are at last liberated from the constraints of cowardice bred of ancient fears, but in which authentic human aspirations can be realized—if not for all time, at least for the foreseeable future.

Interestingly, Marx proclaimed a form of militant atheism in the nineteenth century. This was the old-time atheism, which he said was purely abstract unless it led to concrete action. This he enshrined in the revolutionary ideology of communism. It became dogmatic and tyrannical; for many of his successors sacrificed the ethics of freedom at the altar of utopian totalitarianism.

In my view, we need to adopt a more realistic approach to the humanist alternative, one that is skeptical of theistic claims yet affirms a genuine and inclusive Neo-Humanism. This emphasizes the importance of individual freedom, human rights, and a good will. It entails the empathetic imperative and the value of joyful creativity and exuberance—and a *new planetary ethics that recognizes the dignity and value of all persons on the planet*. It is my conviction that humankind can continue the progressive amelioration of the human condition and the realization of our highest hopes and aspirations—this in spite of the fact that we live in a turbulent, unpredictable, and often precarious universe.

THE OPERATIC SCENARIO OF THIS WORK

I have labeled the various parts of this work *Overture*, *Intermezzo*, Acts One through Nine, and *Grand Finale* because I wish to emphasize the *dramatic* character of the biosphere, human affairs, and of the physical universe itself. Perhaps the secret of nature is that it is like the *opera*, which is a comic-tragic narrative of human life, in which rationality is infused with passion. The difference of course is that an opera is a fictionalized work of art presenting a spectacle on stage, with scenery, human characters in colorful costumes, a script, lyrics, choral voices, and an orchestra of strings, brass, and percussion instruments accompanying the soprano, coloratura, tenor, baritone, or bass. The opera is a stunning aesthetic rendition of human transactions, whereas nature and life are *real* yet often manifest similar intensity and mystery. From the human perspective, it generates enthusiastic bursts of hope and splendor, tragic outcomes, despair and defeat. To consider nature "operatic" is of course only poetic metaphor. Similar analogies may be found with other kinds of musical compositions, such as the symphony or jazz, which may ape nature. Nature, in all of its richness, displays powerful and dramatic characteristics of both dissonance and harmony.

We live in a beautiful and elegant universe—as viewed from the deepest laws of nature discovered by science.[2] Yet at the same time, we find it to be a universe in which contingency, chance, dissonance, individuation, historicity, emergence, and novelty are ever present. Indeed, some have speculated that our universe may be only one of a continuing series of creative and destructive universes.[3] The terrestrial sphere of human conduct likewise occurs in an uncertain, often chaotic environment. We have been witness to several destructive earthquakes in the early part of the twenty-first century—in the Indian Ocean, Haiti, Chile, and China, to name but a few—that caused hundreds of thousands of deaths and dramatically illustrate this reality. In spite of this, humans are capable of making moral choices, and the future of the human adventure depends in some measure on how we respond to the challenges encountered in a random universe.

Given the proven track record of scientific inquiry, we need to base our knowledge of nature on the methods of science, the basic methodological principle of naturalism. What the sciences tell us today about the universe—as I interpret it—the reader may find unexpected, even surprising. If my account of the generic traits of nature is warranted, this has important implications for humanist *eupraxsophy*, a term I have introduced to characterize a nonreligious way of life. Derived from three Greek roots, *eu* (good, well), *praxis* (practice, conduct), and *sophia* (wisdom), eupraxsophy enables us to make reflective moral choices inspired by scientific wisdom.[4]

ACT ONE
METANATURE

THE EMERGENCE OF SCIENCE

A radical change in understanding nature occurred in history when humans recognized that supernatural explanations could not account for natural phenomena. The earliest pre-Socratic philosophers in ancient Greece sought to explain events by reference to natural causes. They appealed to reason and observation to interpret nature, not faith or revelation, miracles or theology, uncorroborated by objective evidence. Modern science did not develop until the Renaissance. The ancients used reason and common sense (for example, Aristotle observed a lunar eclipse and reasoned that Earth must be a sphere because it cast a round shadow). Modern scientists developed new experimental methods to interpret nature.

Scientists in the modern world have continued to raise intriguing questions about the nature of the universe. They have asked whether it all fits together, and if so, how. And they continue to probe the implications of the scientific outlook for a clearer, more accurate understanding of the human condition. The Copernican Revolution of the fifteenth century achieved a major breakthrough. It placed the Sun, rather than the Earth, at the center of the solar system. The Darwinian theory of evolution in the nineteenth century replaced former doctrines of creation and intelligent design. The difference between present-day cosmologies and those of the historical past is that modern cosmologies are based on the methods of science, and this includes both mathematical coherence in the formulation of theories and the experimental confirmation of their adequacy.

Modern physics and astronomy began by stepping outside religious authority. The medieval church at first opposed the new science, imposing theological constraints on inquiry. In the sixteenth and seventeenth centuries, the natural philosophers (as they were called), including Copernicus, Galileo, Kepler, and Newton, rejected occult causes and developed the laws of mechanics based on careful observations and mathematical precision. These scientific cosmologists depicted the universe as a fixed system governed by universal laws. The model was similar to a clock or machine, in which every cog and wheel is interconnected with every other. Within the whole, the picture was mechanistic and deterministic.

There was great confidence in the power of mathematical rationality coupled with experimental observation to unravel our understanding of the universe. The poet Alexander Pope extolled Newton as such:

> *Nature and nature's laws laid hid in night.*
> *God said, let Newton be!*
> *And all was light.*

If we knew the exact positions and velocities of the material objects within the universe, we could predict with precision the state of all material events in the future, declared the French astronomer and mathematician Pierre-Simon Laplace at the end of the eighteenth century. What is the place of God in the materialistic scheme of things? asked Napoleon Bonaparte. "Sire, I have no need of that hypothesis," Laplace was reputed to have replied.

By the nineteenth century, it was widely believed that Newtonian physics would permit the scientist to understand the total state of mass and energy throughout the universe. It was also believed during the Enlightenment that the natural sciences could be extended beyond physics to chemistry, biology, psychology, and the social sciences. At the beginning of the Industrial Revolution, French philosopher and mathematician Marquis de Condorcet confidentially prophesized that this knowledge would contribute to the progressive improvement of humankind, including free public education, equal rights for women and racial minorities, a constitutional republic, a liberal economy, and democracy. He died in prison, sacrificed by the passions unleashed during the French Revolution.

The revolutionary findings of Charles Darwin, developed during his voyage on the *Beagle* to the Galapagos Islands, gave a rude jolt to the belief of theologians that all species designed by a divine intelligence were fixed and eternal. Instead, the principles of natural selection were presented as an explanation of how species evolved, including *the descent of man*. Evolution had been suggested by Empedocles in the ancient Hellenistic world, though it was rejected by Aristotle. For the first time, science took history seriously by attempting to explain how things change throughout time.

In the eighteenth and nineteenth centuries the social sciences began to develop daring new ideas. Voyages to unexplored continents led to the comparative studies of anthropology and sociology. Works such as Niccolò Machiavelli's *The Prince* offered astute and often ruthless prescriptions for how to seize and hold power. This led to the development of a realistic study of politics and the eventual emergence of political science in the eighteenth century. Adam Smith's influential book *The Wealth of Nations* had sparked political economy, which led to the new science of economics, to which David Ricardo, John Stuart Mill, Karl Marx, and others contributed. The founding of psychological laboratories by William James at Harvard and Wilhelm Wundt at Leipzig raised expectations that we could understand psychological experience objectively by studying behavior. Psychologists have emphasized the need for testable experimental studies. Today many scientists and philosophers believe that neuroscience will be able to chart the microgeography of the brain and thus understand consciousness in objective neurological terms.

In the twentieth century the theory of relativity introduced by Albert Einstein altered classical conceptions of absolute space and time, and quantum mechanics transformed classical physics by postulating the uncertainty principle of Werner Heisenberg. The dramatic findings in atomic and subatomic theory in the twentieth century have altered our conceptions of how nature operates. Is chance a real factor in nature? Is the universe open to contingency, indeterminacy, diversity—no longer a unified or fixed system but full of process and change? Astronomy in the twentieth century has extended our conceptions of the universe and its dimensions. For contemporary astronomers the universe is expanding rapidly. The big bang theory was postulated to explain this. By spectroscopic analysis of light from the stars and galaxies, an observed shift of that light toward the red end of the color spectrum indicates that the speed of this expansion is increasing.

THE GENERIC TRAITS OF NATURE

The best approach to understanding the world of nature is to turn to the sciences, which attempt to explain how and why nature operates the way it does. The division of labor between scientific disciplines, however, has proliferated, and new specialties have appeared at a breathtaking pace. As a consequence, it is difficult to find a unitary theory that will explain everything. Rather, a feasible goal is to develop a set of generic categories drawn from the various sciences, which, at the very least, describes the broad contours of nature.

Any attempt to understand the generic traits of nature is in itself not a simple task, given the rapid growth of separate disciplines; yet we need to attempt this on an interdisciplinary scale. To understand nature we need to draw upon our observations of data by means of meticulous descriptions and measurements. We then need to develop hypotheses and theories to explain how what is observed is happening. Scientists historically describe and classify things and their properties, and they catalog different kinds of objects, events, and processes. But their basic interest is to formulate causal explanations of how the phenomena under observation behave the way they do. They seek to develop theories and endeavor to test them experimentally. To be accepted within the body of knowledge, such theories need to be replicated and corroborated by a community of inquirers committed to the methods of science. The justification of a theory must be open to inspection and peer review. Often the knowledge obtained is piecemeal, and it may be limited to the highly specialized context under inquiry. There are constant efforts to extrapolate what we learn and to apply this knowledge to other domains if possible. Of course, the conceptual development of theories depends upon mathematics, which serves as an essential tool of inquiry. Thus, the theories of mechanics have had a wide range of applications in field after field, and this has contributed greatly to the growth of physics and astronomy.

We also draw on generic presuppositions that may not be formulated in lawlike fashion or tested experimentally, yet they may serve as powerful analogies for a wide range of subject matter. These generic principles serve as guidelines for inquiry in other fields. A good illustration is the atomic theory, which is a comprehensive interpretation of all things encountered in the world of nature. Leucippus and Democritus postulated an atomic theory among the pre-Socratics of ancient Greece, but this was purely speculative and had no experimental basis, nor was it interpreted mathematically. It was not until the dawn of Newtonian science in the seventeenth century that it was applied to physics and confirmed experimentally using the powerful tool of calculus.

The germ theory of disease is another generic principle that cuts across the disciplines of biology and medicine, and serves as a guide for researchers attempting to cope with the causes of infections, and possible therapies. Antonie van Leeuwenhoek invented a microscope in the late 1600s that detected micro-bacteria. Edward Jenner, Joseph Lister, and others sought to discover infectious bacteria and the vaccines to combat them. Louis Pasteur revolutionized medical science earlier by advising the heating of wine and later the pasteurization of milk to kill germs. He dealt with rabies and anthrax. Ignaz Semmelweis attempted to persuade doctors and midwives to use hand-washing to reduce the incidence of fatal infections contracted during birth delivery. Widespread diseases such as tuberculosis seemed incurable until much later, when the bacillus that infected lungs was detected and sulfanilamide and antibiotics were developed to treat the disease.

The impressive progress in the twentieth century in developing the atomic theory fascinated generations, and Nobel Prizes were awarded for these remarkable discoveries in physics. Philosophers have attempted to interpret the implications of the general atomic theory to our conception of nature. The dawn of the atomic age by midcentury provoked a widespread public debate about contamination from nuclear weapons and nuclear power plants. Every breakthrough in subatomic physics intrigued the public. What did this portend for our understanding of nature? Would an application of the basic laws of physics and chemistry—mass and energy on the atomic and subatomic level—suffice for us to understand all phenomena? This was the thesis of physicalist reductionists, which I think is limited. For we may need to introduce still other principles on more complex levels to understand life, psychology, and sociology. We need higher-order concepts and theories to explain organic matter, biological processes in the biosphere, human psychology, and social institutions.

The advances in physics were made possible by the invention of new instruments to extend our powers of observation. First was the invention of the microscope, which clearly showed the existence of entities in the microworld. Later came the electron microscope, and still later the supercollider, which enables researchers to penetrate an opaque subatomic world that had only been inferred and remained hidden, but traces of which could now be detected in the experimental laboratory. Will the invention of new instruments to probe the brain better enable us in the future to understand human cognition and psychology?

Chemistry has made rapid advances by applying the atomic theory. The primary subject matter of chemistry is molecules. The ability of chemists to analyze compounds and synthesize new molecules—from the world of plastics to new metal alloys—is essential for the industrial and technological applications in the contemporary world. Consumers are impressed by the daring new powers that these discoveries offer for the betterment of living. The emergence of a new pharmacological industry also depended in large part on chemistry and biotechnology. Similarly for the invention of new technologies in medicine, including X-rays, CAT scans, MRIs, and other diagnostic technologies. Researchers are now able to confirm the actual existence of microbes, bacteria, and viruses, and to develop vaccines and antibiotics to cure formerly intractable illnesses. Especially exciting in awakening human imagination were new technologies that enlarged our observation of nature. This was made possible at first by the dramatic use of the telescope to challenge received doctrine. Galileo was able to see the moons of Jupiter and thus overthrow the intransigent authority of the Church, which had postulated a fixed order of nature, invoking Aristotle and Aquinas as the final arbiters of scientific-philosophical questions.

The great breakthroughs in astronomy are of incalculable significance today. The "Hubblean Revolution," named after Edwin P. Hubble, the American astronomer, clearly demonstrated in the 1920s that our own galaxy was not the only one in the universe, and that the cloudy blotches seen on photographs taken from telescopes were not part of our Milky Way but were themselves massive galaxies that had first been labeled as "nebulae." These star systems are separated from us by astounding distances. Moreover, they are traveling away from us at enormous velocities, and the more distant the galaxy, the more rapid its velocity. These conjectures are based on the ingenious spectroscopic analysis of the light waves received by our telescopes. The distant galaxies are traveling enormous distances. As contemporary astronomy advanced, it postulated new dimensions of the universe, the birth and death of stars, the collision of galaxies, the existence of dark matter, wormholes and black holes, supernovae, dwarfs, quasars, pulsars, and it discovered new planets in other solar systems. All of this boggles human imagination and has deflated the special status of the human species in the universe and the historic conceit that humankind was at the center of things and had a privileged place in the universe. Indeed, God was fashioned in the image of man. Alas, we have discovered that we are only one species among innumerable others on a minor planet in a modest solar system on the edge of one galaxy of the billions that exist.

What does this mean to our conceptions of the universe and the place of the human species within it? Formerly we focused on our own solar system and galaxy, but the improved Hubble and *Kepler* telescopes opened up our universe to dimensions heretofore only surmised, not observed, at cosmic leaps beyond ours. As our knowledge expands, the human species seems to shrink in proportion, even though the advance of our scientific and philosophical comprehension of the vast scope of things is truly awe-inspiring.

No doubt these telescopes, and new instruments yet to be developed, will catapult the powers of human observation to other worlds and other possible dimensions. These powers were expanded still further by the advent of the Space Age. The launching of satellites into outer space enables our vision to go beyond the distortions of our own atmosphere, not only by sending telescopes into clearer space, but also by shooting spaceships to the Moon and satellites to Mars, Venus, Mercury, Jupiter, Saturn, and other planets, and even by leaving our solar system and transmitting data back. This is truly impressive, yet the more we discover, the more we shrink in significance. The immensity of the universe and the extensive time scales reduce in significance what happens on our planet, which is of lesser importance in the nature of things.

Our galaxy, the Milky Way, has an estimated 400 billion stars, countless planets, and enormous clouds of gas. The spiral arms of the Milky Way extend approximately 50,000 light-years and revolve every 220 million years, speeding through space at 400,000 kilometers per hour, heading toward our nearest galaxy, Andromeda. It appears to be on a collision course with it. There are billions of additional galaxies laden with stars.[1] Is all of this designed with the human species in mind by a divine being whose son is sent in the image of man, and whose prophets describe a heavenly abode created for human beings? This theistic postulate is a form of "delusion," says Richard Dawkins. What presumption! It is sheer anthropomorphic fantasy.

The fact that humans are able to fathom so much about the universe by drawing on science is impressive, at least in comparison with other species on our planet. Thus we need to question the classical and modern premise of philosophy, theology, and science that the universe is an elegant system of fixed laws and a perfect order. This concept has its origin in Plato's conception of ideas, and that of Aristotle and Aquinas, of a fixed order of species; and it was also a postulate of Newtonian science.

The area of scientific inquiry that has forced a radical reconceptualization of the traditional view of the universe is the Darwinian revolution. This has placed the biosphere at the center of human interest of so much of present-day science. There is fascination with Darwin's hypothesis of natural selection, the process by which new species emerge and others become extinct. Modern theories of evolution recognize the vital role of genetic mutations in providing new capacities to members of a species, which are favorable for its survival, and which by means of differential reproduction may be transmitted to offspring. In this way, there is a change of species and the evolution of new ones. Chance and adaptation thus play vital roles in the biosphere as a multiplicity of new species continues to evolve. The extraordinary diversity of the biosphere reveals so much about nature. There are an estimated 10,000 species of fungi, 270,000 types of flowers and plants, and a multifarious range of mammals on land and sea life in the oceans—a huge panoramic kaleidoscope of life forms. Individuation and historicity characterize them as well as harmony and order. Individuation is a brute fact of nature, whether it applies to planets or galaxies, rainforests or caverns, flowers or social institutions. And all things seem to come into being, change, and pass away.

A stark illustration of this is the massive extinctions of species that have been uncovered. A dramatic example of this are the fossils discovered in the Burgess Shale in Canada, most of them from the Cambrian period of five hundred million years ago. It indicates the great diversity of life forms that evolved with intricate anatomical features but that are now extinct. Thousands of specimens have been assembled, most without any resemblance to current organisms. It also demonstrates that no species is eternal—not even our own—because millions of species have become extinct. There have been several species of humans, and all, including the Neanderthals, are now extinct, except for *Homo sapiens*. Is the great illusion of human immortality—the belief that the human species will persist forever—now shattered once and for all? What does this portend for the human quest for eternity? Is there any consolation for humankind now bereft of salvation?

Well, yes, in a dramatic sense, for although we live in an ordered yet indeterminate universe, we have a creative role to play in our own lives. Adaptation, discovery, and innovation imply a creative aspect to the evolutionary process of the human species. We have discovered that the universe manifests order and disorder, diversity and contingency, and especially uncertainty and ambiguity. All of this is amplified in the human domain, which implies an open and free arena for behavior. We live in a restless and unfinished universe, and human affairs manifest all of these characteristics. What are the ethical implications of living in an open universe in which the end purposes are not predetermined, and in which everything we do has consequences? We have some power of choice, however modest, in the *lebenswelt* (life world) that is our own, and we impinge directly on the environment in which we interact.

PHILOSOPHICAL REFLECTIONS

These reflections are drawn from what the natural biological and social sciences tell us about nature and life. What role, if any, does philosophy have in this inquiry? This book is a treatise not only on nature but also on the human condition. Insofar as philosophy has anything to say, it is on the level of *metanature*. This is somewhat analogous to Aristotle's *metaphysics*, which included treatises that were later compiled by the editors of his works. The books on metaphysics were placed after his writings or lecture notes on physics. What is important about Aristotle's corpus is that he generated his philosophical ideas based on the sciences of his day, which are now, of course, largely outdated.

In *The Turbulent Universe*, my reflections are based on what we have discovered about the physical universe, the biosphere, and the human sphere; that is, upon the scientific accounts of nature that are now available. This is drawn from interdisciplinary scientific fields; they rely upon the methods of science for establishing their truth claims. It is a synoptic view of the universe, a cosmic outlook at this stage in the development of human knowledge. It is a *conceptual landscape* of some of the main features of what we know about nature, and it incorporates the human species and human civilizations in the nature of things.

In a more technical sense, I am attempting to develop a set of *basic categories, the generic traits or generalizations of our knowledge about nature and life*. I am seeking to describe the *conceptual framework* and its basic presuppositions. The biologist E. O. Wilson characterizes this as *consilience*, which means the statement of some of the main interdisciplinary principles at work, the results of our knowledge of nature. Wilson borrowed *consilience* from the English scientist and philosopher William Whewell. In his *Philosophy and the Inductive Sciences*, Whewell introduced *consilience*, referring to the unification of knowledge as derived from the different branches of knowledge.[2] In an age when scientists were becoming more specialized, he approached knowledge as a generalist.

A search for descriptive generalizations seeks to find common traits or characteristics across fields, not scientific laws but rather analogical similarities, such as the atomic theory or the germ theory of illness. So the view that nature reveals evidence of chance events and that what will endure is a product of interacting lines of causality seems corroborated by contemporary science. In response to Einstein's famous statement that "God does not play dice with the universe," I maintain that He or She does indeed, and that the universe is like a game of craps in the sense that we never know which numbers are going to come up. In his quote Einstein was referring to randomness inherent in quantum mechanics, a problem that he wrestled with all of his life without solving it. My qualification is that there is insufficient evidence that God had a hand in the universe or even that God exists. But if we interpret Einstein's statement metaphorically, my answer is that God does indeed play dice with the universe in the sense that what will happen is indeterminate, dependent on competing contingencies and probabilities at any one moment in time. It all depends on the initial conditions that are present.

A factorial analysis of events is central to the thesis that there are many lines of causality interacting, and that *coduction*—based on induction, deduction, and abduction—best characterizes an essential component of a contextual, pluralistic, and perspectival naturalism.[3] This outlook accounts for what is happening in many fields of inquiry. It is, I submit, a fair description of what we experience, and it does not seek to explain data away by a seductive reductionism. This coductive approach allows for the existence of novelty, emergent qualities, and unanticipated turning points in nature. Although mathematical concepts and formal statement of scientific laws are an essential part of science, these conceptual statements are hypothetical, and they apply only as abstractions from a real world. They are in the last analysis convenient pragmatic instruments that are useful in advancing the goals of scientific explanation. They do not have, as far as I can tell, any independent ontological status in the nature of things. Although they designate relationships between observed data and are properties or qualities of mass and energy, they do not make any sense if they are viewed as abstract essences with any sort of prior existence.

The world we encounter is populated by microparticles, molecules, and inert objects of various dimensions and scales, including that which is encountered on the planet Earth; in the solar system; and in other solar systems, galaxies, and clusters of galaxies. This includes organic matter on various levels, from bacteria to organisms; it also includes the blood and guts of lived experiences and reflections on birth, senescence, death, and the processes of evolution and change in historical frameworks. It is a pluralistic, multifaceted scene of splendor and decay, of renewal and destruction, the life and death of stars and galaxies, of species and rainforests, of social institutions and cultures. It demonstrates the role of chance and contingency in the spectacle of nature. Accordingly, we may ask, "What are the generic traits of the universe? Can we fathom its meaning and structure, or is it beyond human comprehension?"

My answer is problematic. Yes, we can expand our understanding of the universe, but there are no easy answers. Can we develop a unified theory in which everything is reduced to a limited number of basic laws, rooted in their physical-chemical sources? An ambitious goal, but whether it will ever be attained is difficult to ascertain *a priori*. We appear to live in a pluralistic, multifaceted universe or multiverse, where an interdisciplinary approach is closer to the actual empirical data. Hence a less ambitious perhaps more achievable goal as a halfway house is more realistic in describing and accounting for the generic properties and characteristics discoverable in the multidisciplinary sciences. Here we find both regularity and chaos. The universe is amenable to elegant, comprehensive mathematical interpretations on the basis of which we often make surprisingly precise predictions; yet at the same time it can best be interpreted by historical reconstructions, and by reference to individuation, uniqueness, chance, and novelty, so that what occurs is not fixed but is contingent on a converging series of causal events. All of this suggests that we live in a turbulent, open-ended universe, where whatever will be is not predetermined or necessary but rather is dependent on processes and events that appear to be contingent. We ask, what is the place of human beings in this orderly yet random universe?

These questions are implicit in the world of nature that our prehistoric ancestors encountered, a world of mystery and awe. These are the questions that continue to entrance human interest ever since, the questions that the pre-Socratics posed and that modern scientists have attempted to solve. These are the questions that I will deal with in these meta-reflections on nature and life.

ACT TWO

CONTINGENCY AND CHANCE IN THE BIOSPHERE

SURVIVAL AND REPRODUCTION

Contingency and chance are pervasive features of nature. This is most dramatically displayed in the biosphere, where species compete in a constant struggle to survive. Whether they succeed is contingent on what happens to them and how they adapt. Life is a marvel to behold. It assumes a great multiplicity of forms, from the simplest one-celled and multicelled organisms to insects, plants, fish, crustaceans, animals, and human beings.

If a living thing is to survive unscathed in a precarious world, it is because it has won out among innumerable adversaries not only in a battle among competing species but also in a contest for fertilization and reproduction among individual members of its own species.

Let us start with the conception of a human being. Of the profuse numbers of human spermatozoa ejaculated by the male's penis into the female's vagina, very few make it up the fallopian tubes to find an egg or eggs to penetrate and to begin the process of fertilization and implantation. All but one (or two or more if twins or triplets, etc.) are wasted. Chance plays a pivotal role here.

The union of an ovum (from a female) and a sperm (from a male) creates a conceptus; that is, two haploid cells merge into a single diploid cell, which is called a zygote. This zygote must next attach itself to the uterus or be dislodged. Many do not succeed in this task. Through a delicate process of development, the conceptus becomes an embryo and eventually a fetus. This new entity may be expelled at any time. There may be a miscarriage; indeed, an estimated one-third of all human pregnancies are prematurely and naturally aborted. If firmly implanted, the fetus is nourished by the mother.

Eventually an infant is born. In past times a great number of babies did not survive childbirth or the first years of infancy when they were prone to infection. Children need to be fed and protected for many years by their parents until they can fend for themselves. The young child is educated within the community through puberty and adolescence to adulthood—until it is old enough to enter into the world on its own, pursue an occupation, and raise a family.

What are the odds that of the countless billions of sperm ejaculated by a male and hundreds of eggs discharged by a female in a lifetime (perhaps twelve per year for forty or more years) that an individual will eventually survive pregnancy, reaching infancy and childhood? And what are the odds of eventually making it through the teenage years, with all of the hazards of living? He or she may be struck down at any point by an unexpected contingency—an accident, misadventure, or a disease that may handicap or destroy that individual. Human institutions and medical science have developed, which enable parents to care for and educate their children, though it was not always the case and it is not the case for wild animals.

Nature is fecund in the biological world. There is an overabundant production of spores, sperm, seeds, and eggs—very few of which will survive. Many acorns fall from a tree, and perhaps only one succeeds in becoming a seedling, nourished by water and soil and sunlight until by sheer luck it eventually becomes a young tree and thrives as a mature tree. Many acorns or seeds end up in the stomachs of animals, are trampled upon the forest floor, or are smothered by other trees or plants under the jungle canopy, each seeking to etch out its own existence. The competition for survival can be intensive and brutal. The plant is eaten by the deer; the deer by a predator. The cycle is endlessly repeated; a raging fire in the California forest is ignited by lightning or by a child playing with matches at a campsite; or an intense drought or hurricane may squelch the collective struggle to survive.

The same tale is enacted time and again in the biosphere. The ant is swallowed by various forms of life that find it succulent; the antelope is hunted and torn apart by wolves; yet some survive to reproduce and continue the arduous task of transmitting their genes. Countless individuals of every species have competed and repeated this drama of life. The cosmic scene is indifferent to whether or not any particular member of a species survives. A chance mutation occurs in a plant or animal that enables it to adapt, a slight advantage is given to it over others, and in time, if favorable, its descendents may benefit and proliferate. Eventually a new species may emerge—there are thousands of different fungi and wildflower species, and thousands of different bird, dragonfly, falcon, and centipede species, which can mate only with their own kind.

And so the drama of life plays out its various scenarios in hazardous and uncertain environments—individuals carrying DNA and chromosome codes challenged to adapt and transmit their genetic stocks, gradually changing and evolving. The processes of evolution are ongoing: selective differentiation and reproduction, order and disorder, normal growth and maladaptation, contingency and change; every new mutation in a predator enabling it to better catch prey leads in time to a concomitant capacity in the prey to flee. The leopard pursues the antelope, which endeavors to make a swifter escape.

The Brazilian, Central American, and African rainforests are lush with overgrowth. As a result of millions of years of chance and conflict, evolution and extinction, a multiplicity of life forms coexists in them. I was fortunate to have observed rainforests firsthand. On entering an overgrown forest in Brazil, one is overwhelmed by the rhapsody and cacophony of sounds, an ebullient showering of colors and tones: fertilization, reproduction, birth, growth, maturation, and death. That this flower or vine, plant or fish, exotic tree, mollusk or piranha, dolphin or crocodile, wild cat or boa, eagle or vulture will eke out its own existence in the rainforest surrounding the mighty Amazon is contingent on many factors in the environment. Is it intelligent design by which all forms of life play out bizarre roles in a never-ending drama of sexuality, exuberant growth, survival, and death? Hardly, for both order and disorder pervade the forest, which sustains life forms within it; yet accidental intrusions may snuff out a life at any time.

The pristine state of nature is hardly an idyllic, beneficent scene of peace and tranquility. It is the stage for living things to realize their potentialities but also to encounter death and destruction. Nature is indifferent to what survives, thrives, or dies.

The theme throughout the great variety of forests and glens, snow fasts and open plains on Earth is *uncertainty*. Each form of life finds exultant satisfactions in ingesting and consummation, erotic pleasure in reproduction, suckling, and nurturing. For those who persist, in spite of hazards and dangers in the process of living, nature is overflowing—with opportunities for new forms of life to appear. It is teeming with fecundity. That some particular living thing exists is a gift; better yet, a bounty, a treasure; for so many events can block its way to live, survive, and persist, in spite of all of the obstacles put in its way; and it is quite a feat for a creature, plant, or organism to succeed.

A similar tale is told about human beings. That this sperm impregnates this egg in sexual pleasure while vast numbers of sperm and eggs—the spores and seeds of life—are dissipated and wasted, as a random gun widely burst in a verdant countryside or mountain vastness (or vagina) spraying everything in sight and out of sight to impregnate. Only one individual (or twins or triplets) makes it and is brought to maturation; all others are vanquished on the roulette wheel of chance.

The outcome of the frenzied battle for survival may be a living, growing entity. These organic cells finally become an individual plant, organism, or human being, though it is constantly challenged to adapt and survive. We can discover the remnant fossils of countless generations of intricately adapted species that thrived for a time, then declined, and eventually became extinct.

On the Australian continent some four and half million years ago, kangaroos emerged. Their fossil remains have been uncovered by paleontologists, who have reconstructed their likely paths of evolution. There are now more than fifty species of kangaroos. Australia at one time contained a lush megafauna rainforest, which supported different species of kangaroo, for we can find fossils of giant kangaroo now extinct. The hypothesis is that the forest gave way to spreading eucalyptus trees, which released less moisture into the air. As the air became less humid, some fauna disappeared. In time, new species of kangaroo proliferated, able to hop to different water holes and feeding grounds. Some eighty kangaroo species have been discovered. They drank less water than goats, and thus they thrived at the expense of other species.

Hence, the first principle of life is the gamble on the odds that my species will survive and thrive, genetically different, sharing a similar code with other members of the species. This is all the more true of the individual who manages to survive and carry in their genes the entire history of the species.

The original state of nature extolled by philosophers and poets has its exhilarating qualities: a tender blade of glass flickering in the Sun, the first lilacs to open with their dreamy fragrance, the romp and play of newly born cubs. "Oh, nothing is so beautiful as Spring," said poet Gerard Manley Hopkins.[1] The signs of renewal and regeneration surround us as the quickening of nature is expressed anew every year.

Geoffrey Chaucer evocatively describes this in his "Prologue" to the *Canterbury Tales*:

> *When in April the sweet showers fall*
> *And pierce the drought of March*
> *And bathe the vein and root*
> *Of every plant with such liquor*
> *That genders forth the flowers*[2]

All of this testifies to the natural exultation of living things as spring bursts forth and the surge and renewal of life forms everywhere. As the sunshine lengthens, it awakens a resplendent outpouring of new growth. The sap rises, silently in the core of the tree, heralding a new beginning for the year. Summer brings the full bloom of fragrance and flowers, new leaves, and corn husks, hazelnuts, cherries, and berries, all ripe for the picking and eating, consummating in an abundant harvest. Nature envelops us with sultry and luxuriant days, cool breezes, and gentle rain by night; gusty winds that clear the air; or thunderous clouds that burst to quench the thirsty soil.

In the autumn there is a welcomed nip in the air at dawn. The leaves turn bright, a kaleidoscope of hues, reds and yellows, oranges and purples, and shades of brown as they begin to gently waft to the matted carpet below, providing new nourishment for the beetles and bugs, worms and slugs, plants and trees.

In the winter months in temperate climates, the first snowflakes herald what is to come. Each snowfall is unique, falling to the ground beneath, covering everything in crystalline coatings. How lovely, as everything sparkles in the wintry Sun. The exquisitely beautiful snowflakes silently fall and build mounds of packed ice.

We contemplate the diversity of nature. No two snowflakes are alike, each with its own wings of intricate shapes, shimmering from the clouds, they are no more as they are blotted together and molded into new forms at ground zero. The fortunes of all living things fluctuate with the times. There is rejoicing at the first opening of the buds and leaves. There is also lamentation at their closing and falling—to conscious bystanders at least, if not to the unconscious forms of life.

Yet nature remains vicarious and precarious. For life in the biosphere there is a time of fragrance, budding, frolicking, and enjoying life, and a time of frustration, as adversity and peril await all forms of life. So there is a polarity, the opening phase of birth, maturation, and exuberance in the fullness of life—and its closing phase, possible extinction at any moment, and finally death and in some species the grieving for lost ones, whether silent or evocative.

In 1918 an influenza epidemic struck innocent people throughout the world without warning—women and men, old people and children, beggars and tycoons alike. There were twenty million deaths, killings at random. The British vessel *Lusitania* sank in the North Sea and great numbers of victims perished; and then there were the casualties of a German submarine in World War I. The great blizzard of 1977 struck Western New York as bitter-cold winds howled across already frozen and snow-covered Lake Erie and dumped huge piles of snow. It was as if it snowed for forty days and forty nights. Snow piled high—up to eighteen feet because of snowplows pushing it to the sides of the roads and the gusty winds adding to its height. The sparrows and crows, wild turkeys and geese, deer and squirrels had a hard time, unable to scratch the ground to find food or nestle in the trees.

Although each life is a paean to nature, each death marks a deafening defeat; all are held hostage to a contingent biosphere. Who will survive in the battle for life depends on how bleak the events surrounding it become.

I reiterate, it is of no concern to nature *at large* what lives or what dies, because there are no discernible fixed purposes in the order of nature. Whether a human being or zebra, mountain goat or beaver is able to survive is a matter of chance and contingency, adaptation and modification. But it matters enormously to those whose lives are the central drama played out on a moving stage.

There is of course discernible regularity and order, and all members of the same species repeat similar processes of birth, development, senescence, and death. There are in-built principles of adaptation and innovation, which enable any one species and its individual members to survive. There is no evidence that nature has been designed by divine providence; there is no justification for believing in an *a priori* blueprint fixed for all time. This is only wishful thinking: but not every wish is father to the fact. There *are* genetic codes that enable the individuals of a species to survive, and these are transmitted to offspring—but there are many that fail. Darwin postulated the process of natural selection to explain how some species emerge and survive while others become extinct. Cognition and intelligence manifests itself in human behavior as a tool of survival, and courage protects fragile human existence from the vicissitudes of fortune. Humans can imitate nature, observed Aristotle, in order to survive by building shelters, lighting fires, gathering and hoarding food, and protecting themselves with crude weaponry. They have invented culture to transmit the tools and knowledge of survival.

HOMEOSTASIS AND THE IMMUNE SYSTEM

Every organism encounters a continuous array of threats to its existence. Among these are pathogens (bacteria, viruses, etc.), which invade the body of the organism and cause disease. Organisms have evolved immune systems that enable them to respond to foreign pathogens. If they are to survive, they must ward off infection and trauma. Thus, mammals have developed effective arsenals of molecules and cells whose specialized task it is to fight the invading pathogens. The first line of defense is this built-in innate immunity. Without it the mammal would succumb to infection. White blood cells (or leukocytes) defend the body against infectious disease. Diverse leukocytes are produced in the bone marrow. The white blood cells exist throughout the body, including the blood and lymphatic system. An increase of the white blood cell count is an indication of infection.

If the organism is bit or cut by a foreign object, various germs may enter the wound, which becomes inflamed and festers. Hopefully the intruding pathogens can be eliminated and those organisms with an effective immune system can be restored to equilibrium and health. Adaptive cells and molecules can come to the rescue to save the integrity of the mammal. If this fails, the result may mean morbidity or death. Perhaps the best illustration for humans is a respiratory infection, such as pneumonia or influenza. The person becomes sick, may become nauseous, tired, and develop fever, signs that the internal mechanisms are at work. If the infection is defeated, the body adapts to the threat.

Unlike other species, humans have developed medicines to help them survive. Thus an antibiotic may be used to thwart a bacterial infection, or a vaccine to prevent a viral infection, or a surgeon's scalpel to rebuild an organ. Humans have actually succeeded in eliminating the scourge of smallpox and other diseases. Unfortunately, pathogens themselves evolve, which may confound the ability of our medicines and our immune systems to cope with them. So there are always new challenges to our health and survival.

Fortunately, natural homeostatic processes operate in all forms of organic life. Organisms need to maintain an internal environment, which, if depleted by dehydration or starvation, can lead to disequilibrium, illness, or death. Functional structures have evolved to enable organisms to survive and to transmit these genetic capacities to similar progeny of their kind. These are not based on a preconceived blueprint, designed by an intelligent designer, but are rather *a posteriori* genetic capacities developed after the fact. We can unravel how new characteristics emerged that enabled new species to evolve. These were due to chance mutations that were favorable for survival. Those species that lacked these mutations disappeared; those that had them could adapt and transmit these characteristics to their offspring. The dinosaur, saber-toothed tiger, and mammoth were unable to adapt and are now extinct, but new forms of life have evolved to inhabit the Earth and to continue "the great chain of *becoming*"—to paraphrase Arthur Lovejoy's "great chain of being." Nature is a dynamic scene, an unfinished, creative, harmonious, destructive, turbulent universe—all this and more. It is not necessarily a unitary, elegant system of fixed scientific laws; the future is not stored in the womb of being. How simplistic is the quest for a unitary mathematical model of all that is or will ever be.[3] Nature is in continuing processes of becoming; it involves symmetry and normalcy but also ruptures, disruptions, and turning points; and this especially applies to the biosphere.

At one time, large portions of the Earth were overflowing with forests and jungles, and life was plentiful. The Earth has been continuously transformed by climatic, geologic, and tectonic changes. Today there are portions of the globe where the climate is severe—extremes in temperature—such as the Sahara or Gobi Deserts; or frigid areas, the North and South Poles and Siberia. There have been dramatic shifts, such as the Ice Age, which overtook large sections of the Earth, or periods of global warming.

THE AMAZON RAINFOREST: PLENITUDE AND WASTE

Where there is natural abundance, life thrives. The rainforests on Earth are especially hospitable to various life forms, each of which endeavors to find a way to survive, and even to exult and luxuriate. The Amazon rainforest is one of the few natural resources of its kind still remaining on Earth. Unfortunately, it is rapidly being degraded by agricultural cultivation and commercial despoliation. Deforestation is the critical first step in destroying this natural scene of the pristine forests, which are rapidly being depleted. The Amazon basin includes parts of Brazil, Colombia, Peru, Venezuela, Ecuador, Guyana, and Suriname. It produces 20 percent of the Earth's oxygen, a vital replenishing source for the planet. What is astonishing is the number of species that live in the Amazon rainforest, many that cannot be found anywhere else on the planet. There are thousands of species of birds in evidence, such as the macaws, which divide into many subgroups: scarlet macaws, blue and gold macaws, chestnut front macaws, macaw clay licks, and more. Similarly for the beautiful blue-headed parrots, or parrots plumed with red and green feathers. There are many tiny hummingbirds, which hover over flowers from which they extract nectar. Other birds can be seen, such as silver-beaked tanagers, blue-crowned marmots, manikins, and jacanas.

In the waters of the Amazon River are yellow piranhas, pink dolphins, cayman (small crocodiles), giant river turtles, and lizards. There are an estimated three hundred species of mammal in the forest: there are many kinds of monkeys—small spider monkeys that leap from tree to tree, wooly squirrel, red howler, and capuchin monkeys. Among the other mammals are sloths and many species of bats and rodents. The capybara is a pig-sized aquatic rodent. There are also tiny deer and giant otters, as well as jaguars, ocelots, pumas, cougars, anteaters, raccoons, tapirs, and armadillos.

The Amazon is home to thousands of varieties of insects, black and brown tarantulas, bullet and fire ants, and many varieties of butterflies flitting among the orchids, lilies, lianas, ferns, acacias, and water plants. There are also carnivorous plants—the Venus flytrap, purple-tipped or Queen's Tears flowers, Poinciana, and other exotic species. The great diversity of life forms is incredible; each has etched itself a specific niche in the forest.

Similarly, a variety of trees proliferate. Many of them tower above to cover the forest with a canopy; other trees are smaller in status, struggling to survive, including the kapok tree, Brazil nut tree, fig tree, and capirona; and many are covered by winding and flowering vines.

Trees are the basis of the forest. There are an estimated fifty thousand varieties of woody plants, trees, shrubs, and vines. The trees take in the sunlight and exude oxygen. On their leafy crowns one hears the calls of different birds that forage fruit and seeds. Boughs, a hundred feet or more above the ground, support abundant epiphytes. These are organisms that attach themselves to living plants or trees. Tall trees create forests high up, which offer microhabitats for creatures that spend their entire lives upstairs: insects, birds, tree frogs, lizards, and many as yet unidentified species.

The overwhelming impression of the rainforest is of *plenitude*—an ecosystem chock-full of varieties of life forms, set in a rich, lush forest, often with high humidity. It is a veritable tangle of intertwined competing forms of organic life, a melee of exotic entities—all part of an intertwined jungle of discordant yet harmonious plants, flowers, trees, insects, birds, fish, and mammals! How sad that much of this ecosystem is now disappearing.

At one time the Amazon rainforest was home to thousands of native tribes; most of them have disappeared or have been torn asunder by colonialization, war, disease, alcohol, or forced labor. For centuries, especially after the Spanish and Portuguese invaders had arrived, the rainforest had offered a shelter for the indigenous population. It provided food and sustenance for the tribes, many of them still naked or covered only with decorative coloring.

The Amazon River basin has a long history of human settlements. Before the arrival of Western colonialists from Europe, complex societies cleared parts of the forest and developed agriculture. They also engaged in fishing. After the colonial invasion an estimated 90 percent died of diseases because of a lack of immunity, and the areas that had been cleared again became overgrown due to neglect or lack of use.

Today there are only an estimated 200 indigenous groups left in the Amazon rainforest, speaking approximately 180 different dialects of 30 language families. The cultural diversity is still very great, and many that live deep within the forest are still relatively isolated. The sizes of these groups range from two hundred to tens of thousands. One can observe the different ethnicities that have developed, particularly in the northwest Amazon. The key historical factor here seems to have been isolation and separation, which in time contributed to different lines of ethnic and social development. Given the fact that hunters and gatherers could find sufficient food to survive meant that there was no need for some of these tribes to develop advanced technological tools or even written languages, though many developed agriculture.

What may we conclude thus far from this foray into a primitive habitat on Earth—the rainforest?

- There is a *plenitude and waste* of life forms in a mosaic intertwined in an ecosystem.
- There is *pluralistic biodiversity* of what is still encountered in the rainforest.
- Here we find a *struggle* of each form of life to find its niche, to receive sunlight, water, moisture, oxygen, and food, and to procreate.
- There is *no discernible master plan*, but continuing transactions in which continuity and discontinuity, regularity, contingency, survival and replication, accident, and death occur.
- Humans are transforming nature all over the planet, often in destructive ways.

This may be generalized to other habitats on Earth. The forests of North America were in pristine form when Europeans arrived to colonize; cut down bushes and trees; till the soil; pursue agriculture; raise cattle and fowl; forge new trails; and build canals, roads, railways, cities, dams, tunnels, and reservoirs. All of this transformed the environment to serve human needs and purposes. This has destroyed the original state of nature. There are other geographical zones on the planet where life has assumed different forms; and there were alternative climatic challenges to which life forms had to adapt if they were to survive. Humankind has altered the natural environment wherever it lives.

In the torrid or tropical zones of the planet, the Sun beats down year-round. The center of this vast area is the equator. Here there are two seasons every year, the wet season and the dry season. Either extreme can be uncomfortable for various forms of life. The Torrid Zone includes northern and central Latin America, northern Australia, Indonesia, Asia, southern India, and Africa.

The virtual opposite of this are the frigid regions of the Earth, the Arctic and Antarctic Circles, the polar zones. These areas are the coldest parts of the Earth and are covered with ice and snow. The Arctic Ocean encircles Alaska, northern Canada, Greenland, northern Scandinavia, and Russia; the Antarctic region includes Antarctica and the southern tips of Argentina and Chile.

The northern temperate zone is between the Arctic Circle and the Tropic of Cancer. This zone includes North America and northern Mexico; northern Asia, Europe, and the British Isles. The southern temperate zone comprises southern Latin American, South Africa, New Zealand, and Australia.

The point is that life has always had to adapt to the geographical and climatic conditions that prevailed in each region. Various parts of the planet are covered by forests, while others are covered by plains. Some areas are dry and desertlike (such as the Sahara and Gobi Deserts); still others are wet and swampy. Some regions are overflowing with rivers and lakes, and others have dry soil or sand.

The truism is that various life forms have had to adapt to such areas if they were to survive. There are millions of life forms that were unable to survive. On the geographic level, this applies to the land masses, whether continents or islands, as the tectonic plates of the Earth's crust shift and move over eons of time. Similarly for the oceans of the planet as well as its lakes and rivers and the multiplicity of aquatic life forms they support.

This brief geographical tour vividly illustrates the role of contingency and chance in the biosphere and the precarious conditions of life forms. In particular the discovery of extinct species graphically demonstrates the reality of death, not only of the individual but of entire species. And it is replete with the drama of uncovering complex and beautiful species that are long-since extinct, as well as those that have recently disappeared and those that are now endangered. Any species may become extinct at any time.

EXTINCTIONS

Five periods of major extinction of life have been identified:

- Late Ordovician period (approximately 445 million years ago)
- Late Devonian period (approximately 370 million years ago)
- End Permian (approximately 250 million years ago)
- End Triassic (approximately 200 million years ago)
- End Cretaceous (approximately 65 million years ago)[4]

E. O. Wilson maintains that the largest extinction in history is occurring due to the effects of human habitation on the planet, especially the growth of population and the exploitation of the planet's resources.

If we go back to the Cambrian period (from 543 to 505 million years ago), we discover the first marine invertebrates, the remains of which can be seen exposed in mountainous lakes and shale. The Burgess Shale is located in Yoho National Park in the Rocky Mountains near British Columbia. It is a rich source of fossils from a period when invertebrate animals, which lack a spinal column, were deposited. This area was probably near the equator at that time. The Burgess Shale contains numerous organisms that have been extremely well preserved when compressed into shale. It is conjectured that the organisms had been transported to this location by mudflows that oozed down the edge of a cliff from a high escarpment. The fossils of organisms and fauna are superbly preserved and abundant in number. Among the specimens are *marrella splendens*, small soft-bodied organisms. These fossils appear as highly compressed films on the rock surface with some of their intricate three-dimensional structures still showing. Other specimens are the *olenoides serratus*, the largest species of trilobites found in the shale, with the soft appendages as well preserved as their skeletons. There are also *sponges*; *tuzoia*, a "bivalved" crustacean similar to modern brine shrimp; and *ottoia*, wormlike creatures with muscle bands and small hooks at one end of the worm. All of these findings have enabled scientists to classify the multiplicity of early life forms that have since disappeared.

The fossils uncovered from the Triassic period are especially dazzling. We can view the huge reconstructed dinosaurs that are on display in science museums.

We can trace the first appearance of the dinosaurs in the Triassic and Jurassic periods when feathered birds and giant plant eaters are seen. The Cretaceous period is the last phase of the age of dinosaurs. Their extinction is postulated to have occurred during the Tertiary period when the dinosaurs disappeared and mammals and birds developed as the dominant species.

Why the dinosaurs became extinct is open to controversy. One theory is that their disappearance took place sixty-five million years ago when an estimated one-third of the Earth's plant and animal species became extinct. Scientists who have examined soil, rocks, and fossils have postulated that something disastrous occurred. One hypothesis is that an asteroid or comet impacted the Earth and that the polluted haze led to the death of plant life (which provided the food supply) and to the mass extinction of dinosaurs. This apparently happened during the Cretaceous and Tertiary geological periods. Others have suggested that the Earth was bombarded for hundreds of thousands of years by asteroids and comets. Some have said that their extinction occurred during various periods of time and had many causes, including volcanic eruptions, microbes, and climatic changes. There is no scientific consensus on the causes, but whatever they are would involve contingent historical events that caused their extinction. Although there were evidently trillions of dinosaur footprints, there are very few traces in rock or shale that have survived. What is interesting is that there were many species of dinosaurs, and that they came in varieties of sizes and dimensions. Evolutionary processes provided a framework for this to happen. This entailed a probabilistic scenario where chance played a role in the processes of natural selection.

The wide diversity of prehistoric dinosaur species—as of all other species—is clearly demonstrated by the fossil record. Some paleontologists have said that during the late Triassic period only forty known genera existed, though the number increased rapidly during the Cretaceous period when at least 245 dinosaur genera lived. During their 160 million years on Earth, their increased diversity apparently occurred when they entered new habitats. They were forced to adapt by becoming more specialized. The full historical reconstruction is of course limited by the geological fossil record that has been uncovered by paleontologists. Obviously many pieces of this jigsaw puzzle have been lost.

The word *dinosaur* was coined in 1842 by British scientist Sir Richard Owen and is derived from the Greek roots *deinos* meaning "marvelous" or "terrible," and *sauros* meaning "lizard." Dinosaurs are distinguished from other prehistoric creatures by their upright legs and the existence of three or more vertebrae that support the pelvic or hip bones. They are classified as reptiles. Dinosaurs have been discovered on various parts of the Earth, the climate of which was different then since there was more carbon dioxide in the atmosphere, and a so-called greenhouse effect existed. The patterns of oceans and continents were markedly different as well: the continents were fused into a huge supercontinent, and the oceans were part of a vast global ocean. The movement of the Earth's crust over time slowly separated into northern and southern continents, and continued to change. The metabolism of dinosaurs was apparently slower to adapt than that of other life forms today.

Many different kinds of dinosaur skeletal remains have been unearthed. The dinosaurs discovered in one area of Zigong, China, show the rich diversity. They include the *Shunosaurus lii* (a medium-sized creature about 9–12 meters in length); the *Omeisaurus tianfuensis* (20 meters in length); the *Yandusaurus multidens* (a small dinosaur of 1.4 meters), which walked on two feet; and the *Mamenchisaurus youngi* (a large dinosaur 16 meters in length).

A great number of extinct animals during this period have been cataloged. There is a long list of about one thousand, including dinosaurs and pterosaurs from A to Z—abdallahsaurus, brachiosaurus, and brontosaurus to zupaysaurus. There also were huge birdlike animals that either are extinct or are the genealogical predecessors of modern birds. Similarly there are lists of aquatic and marine reptiles known as plesiosaurs and ichthyosaurs. New fossils are still being uncovered, such as monstrous pliosaurs discovered in 2009 on a Norwegian island some eight hundred miles from the North Pole. This goliath moved under water with four mighty flippers and preyed upon fish, squid-like creatures, and other reptiles. It had an elongated skull 10 feet long with crushing jaws. The extinct reptile was 50 feet long and weighed 45 tons. It apparently lived during the Jurassic period. It apparently was not directly related to land-roaming dinosaurs.[4]

The fossil remains of the great dinosaurs are mute testimony to the millions of years that these creatures roamed the planet, fed and battled, reproduced and died—and today are no more. What a magnificent drama—the beginning and end of an entire form of life. These creatures were replaced by birds and mammals that appeared on Earth later and repeated similar sequences of events in their evolutionary processes, though many of their descendents still exist. Viewing the huge skeletal reconstructions of the tyrannosaur in natural history museums excites wonder and horror. Such creatures are depicted in a novel by Michael Crichton, which later became the movie *Jurassic Park*. All of this is testimony to the role that chance and contingency plays in the biosphere, in both the past and the present.

It would be instructive to illustrate the numbers of species that have become extinct in the more recent Holocene epoch (the present period that can be traced back approximately 11,700 years). Much of this epoch was concurrent with the decline of the Ice Age and the rise of human civilization. With the melting of glaciers, sea levels began to rise. This permitted marine life incursions over a period of time in many areas not previously covered by water. Surprisingly, fossil remains have been uncovered in many areas far inland. The global climate became warmer, though apparently there were cyclical periods of cooling and warming. This made it possible for habitation by various species to move northward (in the Northern Hemisphere) as the glaciers receded.

These climatic and geographic changes had a direct effect upon the distribution of plant and animal life. As a result many large animals, such as mastodons, mammoths, saber-toothed tigers, and giant sloth, became extinct during glacialization. Other events, such as intermittent asteroid and meteor showers, earthquakes, and volcanoes, also affected various forms of life on the planet. Life, already precarious in certain periods, became even more threatened.

It is interesting to outline some of the animals that became extinct in South America, notably in the Amazon rainforest, which I earlier described. Some of these extinctions are rather recent, based on the testimony of individuals living in the areas these animals once inhabited. For example, *Phrynomedusa fimbriata* is a species of Brazilian frog that is no longer seen; similarly for Darwin's rice rat, until recently it inhabited Ecuador. Extinct birds include the Colombian grebe, the red-throated woodrat, and the niceforo brown pintail duck. Similar mortuary lists can be drawn up for amphibians, mollusks, and insects. I refer to these recent extinctions only to reiterate the thesis that the natural habitat of life forms and the battle for survival is ubiquitous; that most—indeed all—life forms that have persisted for periods of time were eventually vanquished because they were unable to adapt. All of this is a stark reminder that nature in its original state was not a place of peace, order, and tranquility. It is overwhelmingly the case that although various forms of life were able to proliferate and luxuriate on the planet during various periods of time, in the long run they all died out. This applies not only to individual members of a species but also to species lines themselves, which were replaced by or evolved into other life forms that were better able to survive. But their reprieve, likewise, was only temporary, perhaps lasting millions of years, but it was not permanent. One can point to various forms of turtles and reptiles, which have survived for long periods of time; but they are surely not eternal and will succumb in time.

If we tour the globe, we can compile extensive extinction lists. In Europe, these would include cave bears and cave lions, dwarf elephants, hippopotamuses, giant rats, and woolly rhinoceroses. These were species that disappeared in the Pleistocene epoch from 1.8 million to 11,700 years ago, before the first rise of extensive human habitation. The decline of innumerable species during the Pleistocene period was also related to intermittent ice ages due to massive solar shifts that affected the planet. The major cause of extinction in the Holocene era, however, was the rise of human civilization as it spread across Europe and other parts of the globe. The rate of extinction has been massive whenever humans dominate an area and transform the natural ecology. It is sometimes stated that in France no bird, rabbit, or boar was safe from the palate of gourmet chefs! In any case, virtually all wild animals have either been hunted down or domesticated.

During the Pleistocene epoch each advance of the glaciers locked in enormous quantities of water, and each period of recession transformed the habitation of life, with plants and trees sprouting up as soon as warming permitted, and various mammals following to consume the new growth. Or conversely, the result was decimation of populations whenever sudden icing overtook the terrain.

The extinctions that occurred during the Holocene period included the Caspian tiger, Caucasian moose, European ass, lion, leopard, lynx, lizard, and reptiles. There were local extinctions in certain areas, if not the destruction of entire species, such as the Anatolian leopard on the island of Samos, the Asiatic lion on the Caucasus Mountains, the gray whale in the North and Irish Seas, the cheetah in Armenia, the great auk bird in Iceland, and the Ibiza ram in Spain. Similarly, there were mass extinctions on the African continent first during the

Pleistocene period: three-toed horse, giant horned buffalo, and African wolf—all now considered to be exotic creatures. During the Holocene period the blue antelope, pygmy hippopotamus, and Ethiopian amphibious rat also became extinct. The list of birds that are seen no more is very long, from the dodo bird to the broad-billed parrot.

The desertification of portions of Earth has wreaked havoc with a high number of life forms. If there is little or no moisture, plants and animals have great difficulty surviving, although camels and prairie dogs have adapted to the intense heat in some desert areas. Graceful gazelle and antelope species of northern Africa are becoming extinct. The growth of cacti in receding areas in Arizona and Nevada is a wonder to behold. Doing without water for extended periods of time, new blooms and flowers burst forth overnight. It is remarkable how many species of cacti can survive in spite of a severe dearth of water; yet the rate of extinction continues.

There was an extensive loss of prehistoric species in North America during the Holocene period, such as the ancient bison, giant beavers, woolly and pygmy mammoths. More recent extinctions include the American cheetah and stag moose, Caribbean monk seal, and Arizona jaguar. The cougars are rapidly disappearing from the formerly wild west.

There have been a great number of extinctions in Asia: Syrian and Chinese elephants; Bali, Caspian, and Java tigers; the giant fruit bat, Arabian gazelle, giant tree rat, scimitar oryx, Siberian tiger, Japanese sea lion, Chinese river dolphin, Asiatic ostrich, and Yunnan box turtles.

How sad for those who cherish the diversity of life on the planet, which is now squeezed out by human exploitation and domination of the environment. The campaign to preserve parts of the national ecology should have some moral claim on the conscience of humankind.

ACT THREE
ADAPTATION AND EVOLUTION

NATURAL SELECTION

What but the wolf's tooth whittled so fine
The fleet limbs of the antelope?
What but fear winged the birds, and hunger
Jeweled with such eyes the great goshawk's head?
Violence has been the sire of all the world's values.

—FROM "THE BLOODY SIRE"
BY ROBINSON JEFFERS[1]

As we have seen, there is extraordinary diversity in the biosphere. How and why this occurred brings us to natural selection and Darwin's theory of evolution. It is estimated that there have been up to thirty million species, of which only two million are known. The total number of extinct species is unknown, though many have suggested that 99 percent of all species that have ever existed are by now extinct. The incredible heterogeneity of life forms boggles our imagination. They run in size from bacteria and *E. coli* to whales and dinosaurs, from boll weevils and spiders to massive redwood trees. They range from life forms at the bottom of the oceans and in hot volcanic areas to fungi and algae existing in the ice masses of the polar caps. There is virtually an infinite variety of species able to cling to life and survive in spite of seemingly impossible odds against them.

Evolutionary biologists have postulated that life on our planet goes back some 3.5 billion years. Most likely it began from a common ancestor, molecule, or cell that was able to divide, ingest nutrients, excrete waste, and reproduce itself. From that primitive life form evolved the immense number of branches and species. Accordingly, there is some genetic connection among all the organisms on Earth, and the history of any species is a process of descent with modifications.[2]

Nevertheless, the extinction of countless species on Earth and the emergence of exotic new ones has been brilliantly explained by natural selection, first introduced by Charles Darwin with the publication of *On the Origin of Species by Means of Natural Selection* in 1859. This is considered to be the most revolutionary principle of biology—and now the cornerstone of all biological and medical science.

Natural selection is the process by which evolution operates. The term *evolution* denotes *the change in the genetic stock of a species through time*. Darwin was himself unaware of the genetic theory that was later developed by the monk Gregor Mendel, who studied the inheritance of traits in pea plants, which eventually became the science of genetics. An integrated theory of evolution was fashioned by the mid-twentieth century. The theory draws on a number of disciplines, including paleontology, ecology, anthropology, biology, molecular biology, and biochemistry.

This new theory overthrew the classical view that species were immutable. This was held by Aristotle, and later supported by theists, because fixed, unchanging species were believed to have been created by God.[3] It was part of intelligent design. According to this belief, all species fulfilled their unique functions within a harmonious order. For example, bats have intricate radar-like apparatuses that enable them to home in on and target insects in the darkness. Hummingbirds can hover in place like helicopters and alter their positions virtually instantaneously; and they have delicately slivered tongues to suck out the nectar hidden within flowers. The flowers attach pollen to their beaks, which enables the hummingbirds to fertilize other flowers as they flit about. They serve as a transmission conduit that makes reproduction possible. Bees are also essential for pollenizing plant life. Moving from plant to plant, the pollen on their heads and bodies impregnates other plants. Without them, agriculture would collapse. The honey the bees produce in turn is consumed by humans, who find the taste succulent. The interdependence of some species on others points to coevolution; that is, that two or more species cooperate and that the evolution of one is tied to that of another, satisfying mutual needs and adapting to the needs of the other. Yet there is also decimation and disaster throughout nature. The recent infestation and decimation of bee populations in the world is a source of great concern because of its rippling effect on various forms of life.

Everything, according to the argument from design, had a function within a well-designed ecosystem. But there are egregious gaps in this alleged perfect order; for a hailstorm too early in the spring may kill young flowers, and an entire line of antelope may become extinct because of predators or hunters. So the natural scheme is not idyllic but rather is overladen with conflict and disharmony.

In examining the bones of various life forms, paleontologists concluded that many were from earlier extinct species; and as they dug deeper into different layers and strata, they found that these bones were not identical. This demonstrated that gradual changes had occurred. Scientists are able to piece together teeth, jaws, skulls, and bones, and they could date them by using radioactive dating, such as carbon-14. This is helpful in determining the age of archaeological artifacts of biological origin, such as bone, wood, cloth, or plant fibers created by humans. Thus it can date a mummy in ancient Egypt, a wooden tool used by primitive people, or drawings made with charcoal on the walls of caves. Carbon-14 decays and is not replaced, whereas carbon-12 remains constant in the sample. Carbon-14 dating is invaluable in examining the sites where fossils are deposited. This depicts the lineage of certain classes of fossils through time, such as the evolution of horses through various stages or even of birds emerging from extinct dinosaurs by developing feathers and the capacity to fly. When threatened, a species must adapt, or it risks becoming extinct.

In *On the Origin of Species*, Charles Darwin stated:

> As more individuals are produced than can possibly survive, there must in every case be a struggle for existence, either one individual with another of the same species, or with the individuals of distinct species, or with the physical conditions of life. . . . Can it, then, be thought improbable, seeing that variations useful to man have undoubtedly occurred, that other variations useful in some way to each being in the great and complex battle of life, should occur

> in the course of many successive generations? If such do occur, can we doubt (remembering that many more individuals are born than can possibly survive) that individuals having any advantage, however slight, over others, would have the best chance of surviving and of procreating their kind? On the other hand, we may feel sure that any variation in the least degree injurious would be rigidly destroyed. This preservation of favourable individual differences and variations, and the destruction of those which are injurious, I have called Natural Selection.[4]

Natural selection was thus proposed by Darwin to account for the *adaptive organizations of living beings*. This is a process that produces evolutionary change through time and results in *diversity*. The multiplicities of species are not directly promoted by natural selection, though *they are the by-products of adaptation to different environments*.

What this means is that unless organisms can adapt to changes in nature, they will not survive. According to latter-day evolutionary theory, chance mutations occur, and, if favorable, they allow individuals who possess them not only *to survive*, but to transmit such characteristics to their offspring. Those who possess them have a tendency to outbreed other members of a species that do not possess them. A species is defined in part because of its capacity for reproduction with similar members of its kind. Where similar but distinct species do interbreed, they end up with a hybrid such as a jackass, which cannot reproduce itself. Natural selection thus seems to operate because of rules or causes that enable some individuals to survive and reproduce and not others. This applies to all species. It explains their ability to survive because of this capacity. This suggests a *lawlike regularity* of a multiplicity of life forms. There is both *chance* and *order* influencing who will survive and why. This is a *comprehensive generalization* that enables us to connect a virtually infinite series of events in the biosphere across disciplines and subject matters.

The principles of natural selection, like Newton's laws of mechanics or the probabilistic principles of quantum mechanics, are powerful explanations, enabling us to fit together a great variety of singular events. In paleontology, diverse fossils of humans, mammals, plants, and marine life are arranged and interpreted in the light of natural selection. Darwin investigated barnacles and estimated that there are 1,200 species of them. Much like a jigsaw puzzle, we are able to fit the pieces together into a coherent picture of evolutionary change. The striking fact that we encounter is the diversity and variability that life forms take.

A composite theory of evolution developed a century after Darwin includes the following components: (1) a common ancestry, (2) a gradual process, (3) genetic mutations, (4) adaptations, (5) natural selection, and (6) speciation. Although there are general principles that apply to any species, the role of contingency and chance, as I have pointed out, is an important factor in what has occurred or will occur. In any case, the evidence for the general theory of evolution is abundant. I wish to focus on four areas of evidence.

First is the massive fossil record that has been painstakingly assembled and pieced together, using the best dating techniques. Circumstantial evidence enables paleontologists to classify bones under a species, showing how it changed through time. Unfortunately, much of the fossil record has been lost, so we work with what we have, which is substantial enough to provide sufficient evidence for evolution.

Second are the results of selective breeding of crops, plants, and domesticated animals to create valuable hybrids. Humans have thus engaged in genetic engineering and directed the course of evolution for thousands of years. Tangerines and nectarines are the luscious illustrations of this, as are thoroughbred racehorses and dogs—from Great Danes to pugs. Similarly, it is now within the power of human beings by eugenics, artificial insemination, and other techniques to select the characteristics of their children. There is here an effort to develop "designer babies." As cloning and stem cell research develop, further experimentation with human progeny is most likely to be attempted. This raises many fears of a "brave new world" being created. No doubt there are some dangers, but there are positive outcomes in using these techniques. If we could cure diabetes, Parkinson's, or Alzheimer's disease, it would be foolhardy not to do so, and few would object other than on theological grounds. It is also tempting to use such techniques to enhance native intelligence, creativity, and physical agility, if possible. That is not the issue now; I am simply pointing out that these powers indicate that the fact that human engineering can alter a species, including our own, is added evidence for evolution.

A third kind of evidence is the discovery of vestigial organs, such as the human appendix, male teats, or the coccyx (human tailbone), which have no apparent functions. Such vestigial organs are the remnants of formerly functional structures. They are holdovers from earlier ancestral stages in the evolution of a species. Small vestiges of leg bones in whales suggest that they once lived ashore. The fact that whales breathe by means of lungs indicates that they are mammals that evolved from land animals (perhaps descending from an animal like the hippopotamus).

A fourth kind of evidence is the development of antibiotics and vaccines to destroy invading bacteria and viruses, and the rapid adaptation in time of strains of bacteria or viruses to the drugs used to combat them. This compels researchers to come up with new forms of antibiotics that can overcome the resistance of the bacteria. Since bacteria proliferate rapidly, we witness evolution occurring in short periods of time. This can be graphically demonstrated by the effort to find new vaccines to combat new viral infections, such as bird flu or swine flu.

Now is a good time to dispel the popular misconception that a scientific theory is merely a guess. Scientific theories are descriptive explanations of observed reality based upon significant amounts of evidence. So the atomic theory of matter, the germ theory of illness, the gene theory of heredity, the theory of gravity, and the theory of evolution are all, in lay terms, *factual*. When additional evidence calls for it, scientists readily modify theories to fit the new evidence. For example, the germ theory was developed in the 1800s, before electron microscopes and other evidence made clear that the term *germ* was inadequate, because it included viruses, bacteria, and fungi, which require different kinds of treatments. Vaccines, for instance, are effective only against viruses, and antibiotics are effective only against bacteria. Scientists did not try to deny or ignore the newly found facts; they eagerly incorporated the new discoveries into what is still called the germ theory of illness.

The evidence for evolution is so extensive that we can assert that it is not simply a theory but an established *fact*. It is a shame that religious proponents of intelligent design have vociferously contested evolution, in spite of converging lines of evidence. This surely does not deny that there can be differences of *how* evolution has occurred, and new causal factors may still be discovered. The composite theory that has been developed enables us to explain a wide range of data, though obviously science is not cast in stone, and its theories may be refined and modified in the future as new evidence points the way. The salient point I wish to underscore again is the role of contingency in the evolution of life forms. *Chance* intervenes in at least two ways: the need to adapt to contingencies encountered in the environment and the role of chance mutations, which may turn out to be favorable and may become generalized throughout a species. All of this demonstrates the role of chance in the biosphere and, more importantly, in nature in general.

There is an eloquent quotation from the philosopher George Santayana that is appropriate here:

> [N]ature is contingent. An infinite canvas is spread before us on which any world might have been painted. The actuality of existent things is sharpened and the possibilities of things are enlarged. We cease to be surprised or distressed at finding existence unstable and transitory . . . but perhaps all existence is in flux even down to its first principles.[5]

HUMAN EVOLUTION

The intriguing question that we now face is *how* human evolution occurred. Charles Darwin proposed the theory that "man is descended from a hairy, tailed quadruped, probably arboreal in its habits." Darwin thought that all species had a common ancestor. "The sole object of this work," he said in *The Descent of Man*, is to consider "whether man, like every other species is descended from a prehistoric form."[6] We share similar traits with other primates, such as apes and monkeys: opposable thumbs, fingernails, thirty-two teeth, forward-facing eyes, and color vision. Indeed, chimpanzees look remarkably human in many ways. This offended the religious sensibilities of the Victorians of Darwin's period, and it still grates on evangelical fundamentalists today, because it challenges the smug attitude that the human species is a product of creation and is separate and distinct from all the rest of nature.

Yet evidence for descent of humans from other primates is overwhelming. Zoologists and paleontologists classify the human species under the family name of Hominidae, which comprises hominids (now represented only by *Homo sapiens*) and pongids (the great apes, such as gorillas, monkeys, bonobos, chimpanzees, orangutans, etc.). The term *hominid* refers to the family of human species.

One thing evangelical fundamentalists cannot account for in their creationist beliefs is that there have been at least half a dozen species of hominids. Neanderthals, for instance, were a different species of humans than were our Cro-Magnon ancestors.[7]

Darwin did not know about the great number of fossils of humans and apes that have been uncovered since his day, further corroborating the thesis. Based on this research, we may trace the origins of the human species to a common ancestor shared by both humans and apes. We have identified this ancestor as *Australopithecus*, which was intermediate between apes and humans, and which lived somewhere between 5.3 million and 4.2 million years ago. There were several varieties of *Australopithecus*: *afarensis*, *africanus*, *robustus*, and *boisei*. All of these beings became extinct 1.1 million years ago. The discovery of Lucy in Ethiopia by the American anthro-paleontologist, Donald Johanson in 1974 was an exciting find. She was an *Australopithecus afarensis* female of between twenty and thirty years old, who walked upright and existed some 3.18 to 3.2 million years ago. Lucy is considered to be a transitional creature between apes and humans.[8] Surprisingly, there were several hominid species: *Homo erectus*, *Homo habilis*, and *Homo sapiens*. *Homo habilis* existed between 1.5 and 2.4 million years ago. It was named *habilis*, which means "handy man," because of the tools found with its remains. *Homo erectus* lived 1.8 million to 300,000 years ago. *Homo sapiens* (the "wise ones") have been identified as *the* human species. Also identified is *Homo neanderthalensis*, which was first discovered in the Neander Valley, Germany, from which it derived its name; similarly for *Homo heidelbergensis* (found near the German city of Heidelberg). The Neanderthals were husky and short; they used tools and weapons and were found throughout Europe and the Middle East between 230,000 and 30,000 years ago.

Another set of fossils was discovered in 2003 on the island of Flores in Indonesia. Called *Homo floresiensis*, these specimens are small in stature, only three- to four-feet tall; hence they were called "hobbits" (after Tolkien's fictionalized characters). They stood upright and possessed a running gait; they were small-brained yet were able to shape tools. A spirited scientific controversy has ensued about their status. Carbon-14 dating indicated that they were living on Flores as recently as 8,000 to 18,000 years ago. Was this another hominid species, a variant of *Homo sapiens*, or was its diminutive, dwarflike size due to an illness? Apparently they have since become extinct, but the controversy has not been resolved as yet.[9]

Jerry A. Coyne points out that environmental changes in Africa, especially drought, caused the rainforests to become open forests, savannas, and grasslands. The hypothesis is that this resulted in the evolution of bipedalism among hominids. The ability to walk upright permitted them to move from one forest to another; it freed their hands to gather food on the ground and/or to pick fruit from lower branches. Food gatherers eventually became hunters. This led to the invention of weapons to hunt and kill animals, and stone tools to carve and scrape the meat.

Homo sapiens are the only hominids that have managed to survive; all of the other hominids have become extinct. It is unclear as to whether all of the above are separate species. Fossils of *Homo erectus* have been discovered in Africa, Indonesia, and China. Coyne observes that "about 60,000 years ago, every *Homo erectus* population suddenly vanished and was replaced by fossils of 'anatomically modern' *Homo sapiens*."[10]

Similarly, *homo neanderthalensis* "also disappeared." Coyne observes that when he was a student, evolutionary scientists believed that the Neanderthals evolved into *Homo sapiens*. But he says that this is incorrect and "around 28,000 years ago, the Neanderthal fossils vanished." They were replaced by *Homo sapiens* "who apparently replaced every other hominid on Earth."[11] "In other words," he opines, "*Homo sapiens* apparently elbowed out every other hominid on Earth."[12]

There are two theories of what happened. The first is that *Homo erectus* and perhaps *Homo neanderthalensis* evolved into *Homo sapiens*, but this would mean that the gene pool of these other hominids was absorbed by *Homo sapiens* independently in various parts of the world by means of natural selection.[13]

The second theory is that *Homo sapiens* originated in Africa and approximately 50,000 to 60,000 years ago migrated to Europe, the Middle East, Asia, Australia, and Indonesia, eventually crossing the Bering Strait about 12,000 to 15,000 years ago (when it was frozen over) to reach North and South America. This would mean that *Homo sapiens* competed with the Neanderthals and *Homo erectus* for food, and perhaps even killed them off.[14] In the 1970s a fairly radical hypothesis was proposed by paleontologists; namely, that *Homo erectus* and the Neanderthals "were actually two distinct species, not the ancestors of *Homo sapiens*."[15]

The fossils of Neanderthals appear to be different from those of the Cro-Magnons that inhabited Europe in the past one hundred thousand years. Paleoanthropologist Allan Wilson analyzed the mitochondrial cells of *Homo sapiens*. These came only from the mother. He found them similar to the cells of human beings throughout the world today, which indicates a common ancestor migrating from Africa. Hominids have lived in various parts of the world, going back millions of years. Fossils of *Homo neanderthalensis* were found in Europe and the Middle East; *Homo erectus* was found throughout Asia. Thus, there is evidence that *Homo sapiens* were a more recent arrival, having left Africa and taken a journey worldwide only in fairly recent times. *Homo sapiens* would then be a distinct species, unlike other hominids. "Africans and Europeans are on one branch; Neanderthals cluster on a separate branch," points out Carl Zimmer.[16] Neanderthals were sturdy creatures, able to survive the Ice Age. They crafted tools and weapons, were able to hunt, and cared for their sick; similarly for *Homo erectus*. Yet both of these species became extinct, and only *Homo sapiens* were able to survive.

Zimmer asks *why*; and he suggests that a key reason is that they possessed a larger brain, due to new mutations, which endowed them with the ability to fashion more sophisticated tools and invent specialized technologies. They were able to fish, hunt, wear clothing and jewelry, etch the magnificent drawings in the Chauvet Caves of France, and create musical instruments found in the Hohle Fels cave in Ulm in southwestern Germany. They were capable of symbolic representation and in time abstract thinking. *Homo sapiens* developed *language* to communicate with other humans and to transmit their technical and artistic skills to future generations. At some point *culture* emerged; its successive advances were retained in human memory and taught to the young, and *Homo sapiens* began to characterize a social group.

Also important is the evolution of moral principles within social groups. Human beings have survived in part because they depended on the extended family, which became the tribe. It is members of small groups or clans that learned that they needed to establish some rules of the game. There were moral principles, such as loyalty and attachment between parents and children, lovers and friends, brothers and sisters. And so moral rules emerged: I call them the "common moral decencies." This includes empathy and sympathy for members of the tribe and the willingness to defend them when threatened or in dire need. Darwin himself notes that an "advancement in the standard of morality will certainly give an immense advantage to one tribe over another." He observed that a tribe that possesses in a high degree "the spirit of patriotism, fidelity, obedience, courage, and sympathy" and whose members were "ready to aid another, and to sacrifice themselves for the common good, would be victorious over most other tribes; and this would be by natural selection."[17]

Thus, moral norms gradually evolved and became part of the cultural values that provide some cohesion within the social group. In time this may have led to the development of a "moral instinct," as some evolutionary psychologists today postulate.[18] Eventually moral empathy was extended beyond the consanguineous tribe to the village, city, or state, and in time to the whole of humankind. It is the transference of the common moral decencies from the small face-to-face group to the wider community of humankind that was the task of religion and philosophy, which has taken humankind a significant step forward. This was due to the universalization of moral principles that civilization would later develop. But I have jumped ahead of the story of humankind for now.

How does culture transmit ideas, beliefs, and moral values? Richard Dawkins has introduced the concept of *meme*.[19] This is a unit of cultural ideas, symbols, or practices that is transmitted from one mind to others through gestures, signs, speech, or rituals. The Greek term is *mimema*, meaning something that is imitated. Memes are cultural analogies to genes. This concept is very insightful because it shows how ideas or practices can be spread among group(s) of people and how they allow cultural identities to emerge. Some have criticized *memetics* for lacking sufficient empirical precision and confirmation. Granted, but nonetheless, the concept of memes plays an explanatory role in highlighting cultural evolution, the dynamic factor that is added to human evolution, enabling us to transform purely biological evolution, which has developed over long periods of time by natural selection. Hence, the term *coevolution* points to a powerful double factor in human evolution, which at a certain stage in human history becomes *gene-culture evolution*.[20]

What is unique among all living creatures on the planet is the ability of *Homo sapiens* to create tools and instruments, to bend nature to human wishes, and to intervene in the world and change it. Robins build nests and beavers build dams, but that is virtually instinctive; humans create skyscrapers and spaceships to the Moon, and that is truly wondrous. We create new medicines to heal illnesses, new ways to produce food, and new vistas in which our imagination is able to soar. Here creative intelligence is able to function as a coping mechanism. Humans no longer need to depend on instinctive behavior in order to survive, because we can evaluate problematic situations in which we are involved, and we can consciously adapt to conditions or change them. We can modify our behaviors and create new things to enjoy; new worlds, rural or urban environments to live in; and new scientific, aesthetic, and philosophical creations to contemplate and appreciate. All other species are products of natural causes—genetic and environmental. Humans are able to understand the conditions in which we live, and to transform nature to satisfy our wishes and desires. Cognition is thus the most effective tool of adaptation of *Homo sapiens*, and it brings new cultural emergents into existence. No longer dependent on the slow process of natural selection to survive, it can leap ahead and transform nature. Unfortunately, these new powers also enable the human species to wage wars of destruction. Creative human intelligence is the supreme instrument of human beneficence, but, used for malicious purposes, it is also the supreme instrument of planetary destruction. Alas, the quest for peace and harmony is the ongoing challenge to the human primate.

Human beings have constructed complex social institutions over the millennia to achieve their aims and goals. They have transformed planet Earth, forged independent states and military alliances. Intelligence has constructed aqueducts and dams, highways and sewer systems, coastal ports and bridges, agriculture and commerce, urban centers and natural parks, hospitals and schools, religious institutions and universities, music and literature, the arts and sciences, theology and philosophy—all of the goods of cultural expression and fulfillment. And these are the things that we live by and are even willing to die for. We are—and this is unique to *Homo sapiens*—the builders of culture, which we add to the world of nature; and this in its own right defines us and becomes part of the natural world.

This capacity has enabled humans to transcend the constraints of our biological endowment. We can drive motorcycles and automobiles, pilot ocean liners and jet planes; we can communicate instantaneously across the globe and shoot into outer space. We have developed diverse systems of laws and moralities, religious creeds and doctrines of human rights, and oligarchic or democratic governance. Thus, it is culture, not nature, which expresses and enhances the human condition as much as or more than the biological and genetic basis of human nature. The naked human ape is clothed with all of the adornments of sociocultural life. What has evolved is the richness and grandeur of civilizations that continually open new vistas for human realization.

What seems apparent in reviewing the history and evolution of the human species on the planet is that it was fashioned by and responded to contingent events. That *Homo sapiens* were able to succeed over competing species (*Homo habilis*, *Homo erectus*, *Homo neanderthalis*, and others) was perhaps due to chance and luck. We are fortunate that we have come as far as we have and survived, indeed thrived. We have surpassed all other species on the planet. Like the dinosaurs of a previous period, we now dominate the planet. It is ours to possess and use. How things will work out in the end is indeterminate; that the human species will survive in the distant future is uncertain. Creative intelligence is the instrument of human greatness and achievement, but it may also be the source of extreme anxiety and fear, unless we can summon the courage to become what we want and preserve some integrity in spite of an awareness of our ultimate ontological fragility and finitude.

SEXUALITY AND LOVE

Many biological processes are shared by living things. For mammals, respiration, digestion, excretion, and stimulus and response mechanisms are similar in function. Of special interest is the role of sexuality in processes of reproduction. Indeed, sex seems to pervade the biosphere. It is a general structure of virtually all forms of organic life. It is possible for members of a species to reproduce asexually. Bacteria can simply begin the process of dividing without the need for sexual mating. Carl Zimmer points to whiptail lizards, which reproduce without the help of a male.[21] But this is rare, and for the most part all species involve mating between a male, who discharges sperm, and a female, whose egg is thereby impregnated. Thus sexual intercourse is what unites virtually all forms of life; it is an exquisite gift of nature, the most intense form of sensuous pleasure; and for those who fully experience the sexual drive, it makes life worth living!

One can speculate about possible explanations for the evolution of sexuality. The most likely is the advantage that it offers to the offspring of a wider distribution of genes. Asexual reproduction would produce replicas of the adult female with identical genetic stock. Sexual reproduction draws on a much more diverse range of chromosomes, and it also incidentally allows for greater individuation. Genetic diversity grows out of variation, and this drives the evolutionary process.[22] Here there is, as it were, a roll of the genetic dice, for there are billions of possible combinations. Chance again intervenes—that this sperm, carrying various genetic codes, would combine with this egg, carrying different characteristics, is due to circumstances at the *time* of copulation and conception. Carl Zimmer proposes that these recombinants would provide some protection against parasites that are able to invade the same kind of host and destroy them; in the process, chance mutations may occur in an individual, and this may prove to be more successful for survival. It may be more difficult for parasites to invade and destroy diverse individuals than in reproduction among members of the exact same kind. Interestingly, males produce an overabundance of sperm, and, as we have seen, most are wasted and very few succeed in penetrating an egg to begin the process of fertilization—with the gamete now drawing on two sets of chromosomes, intertwined rather than exactly the same set.

It is interesting that sexual attraction begins when ovulation sends pheromones to lure the male with his sperm; and *sexual selection*, according to Darwin, is everywhere present. It begins when males and females begin to attract each other—as soon as the male and female reach puberty. There are two processes at work here: first, there is intense competition among males for the sexual favor of the female, which eventuates in an act of copulation. In some species the battle of males for females can be intense. In some species the male develops horns (as with rhinoceros) or antlers (as with elk) to battle other males for conquest of the female. Sometimes it can be a ferocious battle, and the dominant male (as in gorilla packs) will drive out all competitors. There is the story of two male bulls watching a herd of females at a distance, and the young bull says to the older one, "Look at those gorgeous cows, let's go over and mount one." In response the older and more powerful male drives off the younger one, gallops to the herd of females, and begins to mount them one by one.

Why is sexual passion such a powerful impulse? Obviously it is because the orgasm is a source of intense pleasure, but also because of the perhaps unconscious tendency or instinct to transmit one's own gene stock to sire future offspring. For example, some forms of male birds that are prepared to mate with females have spurs on their penises, which, when penetrating the female, is first able to scrape out any sperm that a previous male may have deposited before he ejaculates his own. Male lions are known to kill the cubs of females that have mated with a different male. Once her cubs are killed, the female goes into estrus, at which time the aggressive lion begins to copulate; and ejaculations may be repeated many times with the female until the male collapses from exhaustion.

Giant walruses will battle other walruses, wound them, and drive them off so that the victor alone can possess the females as his sexual partners. The harem principle applies to many different species, where one male sires a great number of offspring, all of which are heirs to his genetic stock. Incidentally, the same thing has been true of some human cultures, where men of power and wealth are able to amass a harem of women. There are reports of Saudi Arabian princes who have sired great numbers of sons and daughters.

There are of course many different kinds of sexual arrangements. Monogamy has been favored recently in Western culture, though this arrangement is often difficult to maintain, and extramarital affairs and divorce may be the result. Many species are polygamous, polyandrous, or polymorphous perverse, and they are willing and able to have sex with many members of the opposite—and perhaps even the same—sex. Although a dominant male chimpanzee will seek to have the exclusive right to sex with females, many are seen to sneak out and fornicate with younger males. The three *M*s stand out today for harem-like sex, Mormons and Muslims with polygamy, and Monks with wild sex with nuns and altar boys!

In sexual liaisons, it is not simply males battling males in a never-ending contest for sexual power and dominance, but females play a vital role themselves in attracting what they view as a desirable sexual partner. And the males of the species will do what they can to woo females. The classical illustration is the peacock, whose long and colorful tail is displayed to lure the female. Jerry A. Coyne speculates that the amount of effort and energy expended by the peacock, which carries around a huge tail of feathers, would seem to violate natural selection, for it often is a burden for peacocks to escape predators and survive.[23] This phenomenon is found in other species as well. There are *sexual* dimorphisms in the male, such as tails, colors, or songs. Many may very well attract females, but their sexual accoutrements and displays are often heavy baggage in the struggle to survive, which is so essential to evolution. Actually, it is *sexual selection* that is most directly at work in the process of natural selection.

The females also play a vital role by choosing the best male, healthy and handsome, as the father of their progeny. It is clearly the strongest males that emerge in the fight among males for the females. But the females' choice is equally important, and the sexiest displays seem to attract the females that are interested in the sturdiest companion to mate with and that perhaps also will help tend to the chicks or cubs after birth, or will help defend the territorial domain that he stakes out. It is not which potential mate *survives* in the contest among males, but *which one has intercourse and reproduces*. That, in the final analysis, is what counts, and the predilections and/or tendencies of the female play a role in the process. After all, it takes two to tango, and the music and castanets as well as the colorful displays and warbles are important overtures to eventual copulation. In human sexual encounters, it is romantic attraction that seals the deal, or at least when it flowers into love it enhances the liaison. *L'amour, toujour, l'amour* is overladen with the perfume and refined delicacy of romance—the sublime, primal, delirious relationship of a muscular male athlete possessing a Tarzan-like physique with an exquisitely perfumed, lilting lilac female. Of course the act of intercourse may be a one-two-three encounter, as when a female is violently attacked and raped, or even an in-and-out event, without the finery and embroidery of Tristan and Isolde or Romeo and Juliet. Sexual reproduction will ensue whether or not there is foreplay, afterplay, or romance.

But the key point to bear in mind is that sexual attraction and passion may inseminate and produce babies, but they are also the progenitors of love; and to fall in love with another person means that marital arrangement may become long-standing. It is not one-night stands of ecstasy that endure but the bonds of affection and commitment; and out of this may develop—particularly for humans—enduring ties of devotion that remain long after the heat of passion or romantic madness become only embers of an intense flame that, though it is kindled from time to time, no longer engulfs a person in a raging fire.

In particular it is the ability of the female to convince the male to stay with her, to tend to the litter that he has sired: if the love of a man for a woman is unrequited, he needs to convince her to accept his proposal of marriage. Herein is the origin of the family: not simply a pack of cave dwellers living together, hunting and gathering food, but a consanguineous group tied together with the bonds of matrimonial love, which is then showered on their babies as filial piety. A division of labor develops. The male defends his mate and brood from other covetous males or wild animals. He hunts for food; she gathers wood, cooks on the fire, and tends to the hearth. The point is clear: although sexuality is a powerful igniter of passion in coitus, it matures into deeper affection and love, and it needs to be extended to the sisters and brothers that live together and to other members of the clan or tribe.

Thus there are other forms of affection that are nourished and mature—the love of parents for their children, reciprocated by the honor and respect of the children for their parents; the love of sisters and brothers for each other—though often rivals for the affection of their parents and turned off by unfair favoritism; and in time the capacity for friendship of members of the same sex may develop, which may also be a source of affection with sexual overtones, the love of a woman for other women, or of men for each other. The tale has been told many times: humans are able to love individuals no matter what their gender. This transgender bond of affection develops as comradeship and collegiality. These can effloresce and reach heights of moral-social relationships implicitly sexual in undertone, equal in dignity and value to other forms of heterosexual love.

Sexuality is *the* vital component of the evolution of the human species, developing emergent qualities that transcend the instinctive act of copulation and reproduction—it is of intrinsic value in and for itself. The human species was able to survive and thrive first because of an enlarged cerebral cortex and the capacity for creative intelligence; humans are able to understand the world and resolve problems that they are able to overcome. But second is the capacity to appreciate many dimensions of love, rooted in sexual foreplay, intercourse, and afterplay, achieving consummatory qualities affecting the bonds that tie people together and becoming the basis for moral relationships in which humans are able to love other persons *as persons* in and for themselves irrespective of gender, ethnicity, national origin, or circumstances of birth. And it is these two powerful impulses—reason and passion—that are both necessary for the civilizing of our original naked apehood and our eventual efflorescence as humanistic persons, capable of new forms of devotion and commitment and new forms of civilization able at times of tranquility to overcome and supersede the violent, aggressive behavior heretofore displayed in human history.

RETROSPECTIVE AND PROSPECTIVE

Evolutionary science has drawn significant conclusions concerning the status of the human species that I wish to highlight. First, viewed retrospectively, virtually all species in the biosphere that have ever existed have become extinct. This emphasizes the persistent yet ultimate fragility of all forms of life on the planet. Second, this applies to the evolution of *Homo sapiens*. The historical record provides evidence that there were several hominid species that competed.

The fact that *Homo sapiens* survived, whereas all other hominids have disappeared, only points to the hazardous character of human existence. Third, if we project our knowledge of our past to its future, we can only be skeptical. For the generic trait exhibited by all forms of life is uncertainty. True, humans have survived thus far, after taking a tortuous journey, but there is no guarantee that the human species will continue to do so in the future. This reflects the generic trait not only of nature but of all forms of life itself.

Accordingly, the prospective future of *Homo sapiens*—depending on how far into the future one projects—is cast into ultimate doubt. Indeed, today for the first time in human history there is widespread apprehension about the long-range survival of humanity. Not only is it clear that every person will someday die—in spite of efforts by theists to deny individual mortality—but now even more unsettling is the prospective future of our own species and the fact that the entire human adventure at some time in the future will most likely dissipate.

INTERMEZZO

Several questions of direct moral relevance to human pride thus confront us—especially in light of the scientific account of how human evolution occurred, and how *Homo sapiens* won out over other competing species and managed to survive. Surely we can no longer insist that we humans are somehow positioned at the center of the natural world, nor do we have a special purpose to fulfill. We must also admit to the shattering of the historic belief that there is a perfect order in the universe. We do live in an orderly universe of sorts, yet it is turbulent throughout—in which unexpected contingencies constantly crop up, and the best recourse that we humans have is to rely upon ourselves to survive.

If the human species is a product of both natural causes and unexpected contingent events, the former religious conviction that the human species has a special place in the universe is profoundly mistaken. The theory of natural selection surely undermines the former confidence of humans that an intelligent being designed an orderly universe and fine-tuned our planet so that our species could evolve. Actually, the universe is *not* that finely tuned, because accidental events constantly intervene and chance invariably upsets orderly development. Indeed, the unpredictable consequences of our own moral choices are often not foreseen. Nor does the optimistic, naive account of unlimited human potentialities suffice, because haphazard events often intrude to undermine our dreams and hopes and our well-conceived plans to bring them about.

If the human species follows the fate of virtually all other species on the planet, it will eventually become extinct. What does this portend for the human prospect? It is no doubt very difficult to contemplate—even for secular atheists and freethinkers—that there may *not* be an ultimate future for the human species on the planet Earth. Those who have great confidence in science and reason and who believe that we *will* be able to save the planet in time may come to realize the slender basis for that belief.

Indeed, we may ask, how certain can we be that humans will endure in the long run in the light of the contingent and turbulent character of the universe? It is clear that luck has played a key part in the evolutionary history of the human species, and that chance will no doubt continue to intervene in the future. Accordingly, we are uncertain that the human species—let alone our planetary abode—will survive the next few centuries beyond the twenty-second century—as human civilization proceeds to radically alter the planet, transform its surface by deforestation, pollute the seas, and continue unabated industrialization and urbanization. If we reflect on the fact that the human species as we know it has endured at least a thousand centuries (namely a hundred thousand years)—given the current pace of change, we have few guarantees that we will endure the next thousand centuries, let alone three or four.[1] This applies not only to our own species, which may follow the course of other species and become extinct, but also to other forms of life. One day the Earth may become virtually uninhabitable as the destruction of the natural ecology, along with other unforeseen natural causes, makes our planet virtually unlivable, except for insects, worms, viruses, and bacteria—so that it is they who may inherit the Earth.

ESCAPING TO OUTER SPACE

Men and women have always dreamed of utopian futures wherein conflict, death, and extinction are overcome. We know of the fate of efforts to create idyllic utopian societies on Earth and of the often tragic outcomes of revolutions betrayed. For post-postmodern men and women there is the hope that humans will eventually be able to inhabit other planets and other worlds. The *Star Trek* scenario of Gene Roddenberry that the human species can escape planet Earth and populate other habitable planetary bodies within our galaxy is widely held; or better yet, the prospective journey to other star systems of humankind is the belief of science fiction buffs. Unfortunately there is always the problem of the vast distances of space and time to reach other solar systems within our galaxy, or in "nearby" galaxies. This is compounded exponentially when we seek conventional planets (or moons) of habitation with the prerequisite amount of oxygen, water, and sunlight for survival, and a temperature range that humans can endure. Again, science fiction has come to the rescue; namely, that humans will construct spacecraft capable of traveling long distances and solving the problems of energy and fuel to make the trip.

The search for extraterrestrial life is an exciting exploration into the unknown. Are we alone in the universe? One of the most exciting discoveries at the beginning of the twenty-first century is the growing list of planetary/solar systems in the Milky Way galaxy. The question is asked: Are any of these hospitable to life? The initial program of the *Kepler* telescope, which was launched in orbit around our Sun in 2009, is to survey some hundred thousand stars in order to detect possible conditions that might support life similar to our own. This means the finding of water on the exoplanet. To be able to sustain life, it is conjectured that these planets should not suffer extremes in temperature. The question to be raised is whether our planet Earth is a rare event in the universe, or whether there are similar planets encircling other stars. In other words, is our planet a rare random individual planetary body in a bottomless sea of stars? Many of the planets already discussed are huge planets, hot and gaseous giants like Jupiter.

Astronomers have discovered several large exoplanets that seem to resemble Earth. Thus far the smallest planets detected are three times as large as Earth. So we need to search for planets similar to ours with the possibility that life has evolved on them. By measuring the reduction of a sun's light as a planet transits around it in orbit, astronomers are able to calculate the size of the exoplanet, its mass and density, and other characteristics about it.

It is a giant leap forward, however, to ascertain whether life forms emerged from the stardust that feels and behaves like the kind of life we know, and whether these life forms evolved consciousness and built civilizations with which we can communicate. Contact with them would mark an enthralling advance in knowledge. But continued painstaking observations need to proceed before we can assert that there is some probability that some forms of extraterrestrial life exist in other solar systems.

The recent discovery by the Hubble Telescope of two new solar systems near ours is an especially exciting event. The first, named Upsilon Andromedae, is only some forty-four light-years from Earth and is orbited by three planets, one of which is larger than Jupiter. The second solar system, Epsilon Eridani, is approximately 10.5 light-years from Earth (some 63 trillion miles away). Thus far three planets are estimated to orbit this star, which is cooler, smaller, and younger (about 850 million years old) than our Sun (4.5 billion years old). This second nearby solar system has a great gas planet 1.5 times heavier than Jupiter. Three belts of asteroids, including icy rings, are apparently encircling the star. The closest asteroid belt is about 280 million miles from it. This is like our Sun's asteroid ring orbiting between Mars and Jupiter. Another belt is as far out as our Uranus orbiting our Sun. It is conjectured that other planets can perhaps be discovered in orbit. These solar systems suggest that although they are different from our own, there are common principles that we can draw upon to explain their formation.

This kind of research expands our knowledge and opens us up to possible new understandings of our universe. It clearly demonstrates the presence of *individuation* in the solar system and the need for *historical* studies of how stars and their planets and moons were formed. The importance of common categories of uniformity and regularity in such inquiries is evident.

Perhaps the most fascinating question for humans to contemplate is whether extraterrestrial life exists on other planets or solar systems. Perhaps we humans are not alone in the universe. The fascination with the possibility of contacting extraterrestrial forms of life has captivated human imagination; the extensive literature of science fiction dramatizes scenarios for future contact with such worlds. The discovery of traces of water on Mars and even on planets in other solar systems further goads our imagination to find other kinds of life—though they may resemble the familiar forms of life that we encounter on Earth only vaguely, or not at all.

I have myself spent several decades investigating paranormal claims in diverse fields of curiosity; this included the investigation of UFO sightings. Were they extraterrestrial in origin? I was number 007 on the daisy chain of skeptical UFOlogists (up to 0019) set up by veteran sleuth Phil Klass, an expert in space technology, who assiduously studied each UFO case to find plausible scenarios, which avid believers claim to have spotted.[2] This was the team of "CSICOPers," as we were called—special investigators of the Committee for the Scientific Investigation of Claims of the Paranormal (CSICOP). We at first took very seriously the testimony of alleged eyewitnesses and their claims that these were from extraterrestrial spacecraft, directed by some form of intelligent beings. I investigated many true believers who claimed to have seen UFOs, and some even claimed that they were abducted aboard an alien ship and encountered creatures from outer space.

I am now skeptical of these innumerable claims and of the UFO stories that I have investigated. Prosaic explanations can be given. I was especially impressed by the research of skeptics such as astronomer Allan Hendry,[3] who meticulously investigated each and every case, and who, wherever there was sufficient information that could be corroborated, invariably came up with prosaic explanations. Skeptical investigators such as Philip Klass, James Oberg, Robert Sheaffer, and others could not find one hard case of extraterrestrial origin that was irrefutable. When asked by a TV moderator what it would take to convince me that these UFOs are real, I responded with tongue in cheek: "If they flushed a commode in outer space"; that is, a demand for tangible evidence! I think that there is some degree of *probability*, however, that other forms of life exist elsewhere in the vast universe, though they may not take any discernible forms encountered on Earth and may not even have any biochemical resemblance to what exists on our planet. In view of the huge distances, it would take an extraordinary amount of time for any such life forms to reach our planet. In any case, we cannot take as a proven fact that intelligent life from elsewhere has been visiting us until we can confirm that conjecture. I view this fascination with extraterrestrial beings as a new form of *space-age religion* that *satisfies the transcendental temptation of human beings*, spun out of human imagination to quench the thirst for the ultimate meaning of existence. Yet we should not rule out the possibility *a priori*. We could arrange a long-range project of space aficionados, so that if one generation of humans could not make it, successive generations bred for some future habitation could possibly achieve that utopian dream. This now takes on operatic dimensions for the human drama.

The distances to far-flung planetary solar systems are on the order of millions of light-years. It has been suggested that extraterrestrial planets similar to Earth are likely to have formed around one of the nearest stars in the Alpha Centauri system, our closest stellar neighbor. This has been described by astronomers as a possible habitable zone where water can exist on the surface of likely planets. Astronomers have discovered additional solar planets beyond our solar system, and there are probably billions of such bodies out there. It is possible that planets in nearby star systems with planets in orbit around them could be seen through high-powered telescopes and could even be visited one day by exploratory spacecraft. We have detected billions and billions of star systems in the universe, so it is an unlimited field for inquiry.

According to astronomer Pierre Kervella, in a paper published in 2003 in the European journal *Astronomy & Astrophysics*, the nearest known stars that are similar to our Sun are called Alpha Centauri A and B.[4] These two stars are gravitationally *related*. They are bound in a star system and include a third star; these are 4.36 light-years away. Alpha Centauri A has been calculated to be 1,061,000 miles in diameter, and B is estimated to be 748,000 miles across. For comparison, our Sun's diameter is approximately 864,000 miles. The third star is called Proxima Centauri and is a red dwarf. Alpha Centauri A and B orbit each other at a distance of 2.2 million miles. Proxima Centauri orbits the other two stars in a large swath.

The distances are so great that it is highly unlikely that earthlings would be able to visit these stars. Perhaps this scenario of humans escaping aboard a spacecraft may provide some comfort to future generations, but it borders on wishful thinking, not unlike the Christian or Muslim belief in salvation after death and transportation to a heavenly abode. The theistic salvation scenario is clearly a matter of pure belief and is scientifically uncorroborated. But what about the science fiction dream about extraterrestrial space travel? Is it not pure fantasy? The entire scenario can be set to music and choruses can sing *hosanna* in anticipation. This inspires a new religious faith that humanity may someday explore the outer reaches of the universe, and this is no doubt a source of great hope for many. It is a prospective leap into the unknown, to say the least. But there are insuperable technical difficulties with this scenario.

A light-year is calculated by multiplying the speed of light at 186,000 miles per second, times minutes, hours, days, and years; and it is a tremendous distance for future generations to breach. The distance of one light-year is 5,865,696,000 miles. Please take along several extra pairs of socks if you are trekking! Yet some scientists do speculate about faster-than-light interstellar travel by means of traversable wormholes or antigravity propulsion systems. Scientists speculate about radically different forms of space travel, including a way to harvest antimatter in order to shave travel time to the Moon or Mars; others talk about lasers and microwave technology as ways to "beam" people and payloads far distances. Skeptics, of course, are apt to cast extreme doubt that humans can ever approach the speed of light or leap beyond it, and this is still in the realm of science fantasy. Thus far, teleportation, though common in *Star Trek*, is purely conjectural.

Nor is it possible for us to travel backward in time, though Hollywood fiction often dramatizes that scenario. This is all a play of imagination, but it is beyond the realm of realistic probability, especially since organic bodies are material and in time decay upon the death of the person, even though a dead person may be embalmed or mummified. It is highly unlikely that the microconstituents of bodies that lived in the past can be reassembled in the future. So time travel backward seems highly improbable, if not bordering on the impossible. This is the new religion of space travel, geared to replace the old-time religions.

The salient point is that we are uncertain about the human prospect in the long run, because what will be is dependent on unforeseeable contingencies and unpredictable threats to our future existence—if a meteor strikes Earth, if an earthquake and massive tidal waves wipe out significant portions of humanity, if an all-out nuclear war happens, or if some other catastrophic event ensues.

The role of indeterminate brute facticity intervening raises questions about the role of chance in the past and in the future. One may well ask: What were the causal forces and contingent factors that shaped the distant past; retrospective detective work is sometimes difficult to verify, but prognosticating what will happen tomorrow, next year, or centuries into the future is even more uncertain. Thus the contingent universe, fragile, precarious, doubtful, is a fact of nature, life, and culture. Seeing things in their realistic perspective is essential for wisdom—but this may also generate an ultimate hopelessness about the human prospect. Comedian-philosopher Woody Allen said that he had lain awake nights, worrying about what would happen when the Sun cools down several billion years from now. Should one become a nihilist in the face of a bleak future? One would hope not, because we can still live comfortably significant lives without worrying now about what will occur eventually in the remote future. It is similar to the statement that we all know today that each of us will die some day in the future. Meanwhile, we should live each moment fully. What will happen to future generations is an interesting speculative puzzle that we can do little about today. If our great-great-grandparents spent all their time fretting about how their great-great-grandchildren (still unborn) would fare, they would hardly have had time to enjoy the multifarious possibilities of living their lives fully. This is a realistic response to metaphysical angst about the unknown existential future of humankind or our solar system. One recommendation is to forget about it. Do not let it keep you awake at night; try to live fully every day. Meanwhile, we need to examine the status of the physical universe and whether or not its future course is also contingent. From what we have learned about the biosphere, similar contingent factors pervade the physical universe as well.

ACT FOUR
THE PHYSICAL UNIVERSE

"NOTHING BUT"

Let us leapfrog from the biosphere to the natural sciences of physics, chemistry, and astronomy. The first question that I wish to raise is: What is the relationship of these physical sciences to the classical ideal of science, which was to discover universal laws (or a comprehensive theoretical system) under which *all* of the sciences could be subsumed? The model here was to achieve a "unified theory of nature." Under this interpretation certain sciences are considered fundamental, and all the others were said to be deducible from them. The physicalist model considers physics and chemistry to be the basic sciences. This is a kind of monism of mass and energy.

Skeptics have questioned whether the goal is to discover universal laws applicable to *all* parts of the universe, and if so, whether this implies a deterministic scheme in which there is no latitude or flexibility in the universe, no free play, looseness, or exceptions. Does this mean that anything and everything that has ever been or will be known is already implicit in the existing order of things, a Platonic kind of formal structure of pure ideas, or the universal laws of nature?

Skeptics of the classical model ask whether there is genuine novelty and uniqueness in a universe of unending newness, in which clash and conflict, birth and death, implosion and explosion are present throughout, and in which creativity, innovation, adaptation, and adjustment are real features. Is there some room in the universe for uncertainty and chance? An alternative conception of the universe or multiverse that has emerged displays a plurality of open-ended systems in which order and disorder, regularity and irregularity, normal and abnormal, location and dislocation are intrinsic. Is the creativity exhibited in the biosphere by various forms of life endemic to aspects of the universe at large in areas still undiscovered? Is it also discoverable in the physical universe as well? If so, then any unified field theory, however successful in astronomy or quantum mechanics, should not exclude the kinds of complexity and diversity observable in other fields of nature.

No doubt some will say that the above views of a pluralistic and precarious universe are pure speculation, without rhyme or reason, unconfirmed generalizations, in the last analysis a form of poetic metaphor. No, not at all. An open-ended universe seems to me to be warranted, and pluralistic naturalism is based on a fair summing-up of the natural universe that we encounter, drawn from many sciences besides physics and chemistry, an extrapolation from their empirical data.

There are at least two conceptions of "the unity of science": (1) an airtight deductive system in which everything is reducible to a physical-chemical base, or (2) a many-faceted conception of the unity of knowledge, based on generic traits and categories drawn from across the sciences, rather than an internally coherent deductive system. Here the term *consilience* better captures what the sciences present to us at this stage of knowledge. It is based on what we observe evidentially, especially in the light of the enormous complexity found in the biosphere and the human sphere. It is also a result of the expanding horizons of our experience and knowledge, which have been afforded by new inventions that extend our powers of observation and discovery, such as the Hubble Telescope and the Large Hadron Collider.

Viewed methodologically, *coduction* provides a more balanced factorial account of nature than does the universal-deterministic-deductive natural science model alone; for it draws upon explanations from various levels of observation, rather than deducing all of them from basic laws. And it can be used to interpret not only the physical universe but also the life world and human affairs as well. Let me say that a nonreductive *pluralistic naturalism* leaves room for an emergent/evolutionary universe in which creativity and chance are present, grounded on a physical-chemical base, unless we are prepared to allow creativity and chance (quantum mechanics) within the physical-chemical base.

My objection to the "God would not play dice with the universe" thesis of

Albert Einstein is that it does not allow for the multiplicity and complexity of events in nature. Why should physics and astronomy—or the physical components of the natural sciences—provide the sole explanatory model for *all* of the other sciences, including those about the organic world?

Logician-philosopher W. V. Quine was a powerful advocate of physicalism. He was very influential in convincing many philosophers to accept scientific naturalism. This was especially the case for analytic philosophers, who earlier in the twentieth century agreed with Ludwig Wittgenstein that philosophy was *not* one of the natural sciences and that its main task was "the logical clarification of language." Today many analytic philosophers are prepared to go beyond this to a naturalized epistemology based on science.

What did Quine mean by naturalism? It is, he said, "the recognition that it is within science itself, and not in some prior philosophy, that reality is to be identified and described."[1] Or, again, he states: "I admit to naturalism and even glory in it. This means . . . pursuing philosophy rather as a part of one's system of the world, continuous with the rest of science."[2]

Quine's naturalism is thus first and foremost committed to the natural sciences. I accept this approach. For Quine, physics is the best paradigm of scientific inquiry, though he also considered chemistry and biology to be natural sciences. He insisted that there are no metaphysical truths independent of science, and that all scientific knowledge is empirical. He also maintained that there is no "first philosophy" providing the presuppositions of science and derived independently of them. Philosophy is accordingly "continuous with science."[3] Quine himself was a defender of physicalism, especially when he stated that "nothing happens in the world, not the flutter of the eyelid, not the flicker of a thought, without some redistribution or microphysical state."[4] Quine's naturalism was, however, nonreductive. Unfortunately, some scientific physicalists still defend a stronger *nothing but* version.

Quine himself did not consider the human sciences to be genuinely scientific, and I think that he is mistaken in this. The term *natural sciences* needs to be more broadly interpreted without allowing the physicalist criterion to lord over the human sciences, such as psychology, anthropology, sociology, economics, and political science, which seek testable hypotheses, though they may not be entirely deducible from their physicalist components. Some would argue that the variables these human sciences test are not as clearly measurable as are mass, velocity, and the like. Their variables are open to interpretation and that is what makes them less scientific.

Supervenient phenomena such as *consciousness* are no less important than the physical-chemical base, which they qualify or supervene upon. Psychology and cognitive science depend upon the neurological roots of psychological functions, but these higher-order phenomena cannot be "committed to the flames," for they are not physicalist even though they do have an explanatory role to play in scientific inquiry.

To maintain, as Quine did, that natural sciences provide the bedrock of all scientific explanation is questionable—unless one is prepared (as I am) to extend the notion of the natural sciences to include not only biochemistry and biology, but also many of the psychological, behavioral, and social sciences.

Moreover, one's conception of the natural-science model is truncated unless it includes *historical inquiry* and *individuation*, which deal with important components of the world. A further explanation of this will be discussed in Act Six; but let me say for now that by individuation I mean the existence of particular things, fields, or events—atoms and molecules, cells and organs, plants and humans, planets and galaxies—encountered in inquiry. The natural sciences seek universal laws, but these need to be applied within historical contexts and to the particular entities observed, the nature of which can affect the conclusions drawn as a result of the scientific inquiry. Thus we draw upon the general principles of geology in order to explain geological formations observed on Earth and other planets. Here circumstantial evidence is used to reconstruct what might have happened in the specific contexts under inquiry. The surface of the Moon is pockmarked with craters, which resulted from the impact of meteors and asteroids, and from volcanic activity in the remote past. The planet Mars gives evidence of erosion and the likelihood of the presence of water and perhaps even primitive forms of life. Comet Shoemaker-Levy crashed into Jupiter on July 16–22, 1994, after breaking into many fragments during a close encounter with Jupiter two years earlier. These were contingent events, ending the career and orbit of the comet. Similarly, as we shall see, astronomers have observed many explosions of supernovae deep in space. How does this fit in with the classical idea that the universe is a fixed, stable order—when it's anything but that?

The quest for physicalist explanations—reductive *and* nonreductive—is what is at issue. This is primarily a *methodological* goal, which I accept as a fruitful program of inquiry, not a final conclusion. The effort to consider physics and chemistry as *the* preeminent sciences is a presupposition of the physicalists that I think is unwarranted at this stage of scientific inquiry, because other sciences are important in explaining unique objects and events. The physicalist model is based, at least in the area of astronomy, on Newtonian/Einsteinian physics, which has had impressive success in extending the frontiers of knowledge. But it does not say enough about the different kinds of explanations found in the life sciences.

Let us not prejudge the issue. In my view we have insufficient evidence that the *entire* known universe is a unified, interlocking, physicalist field, such that physical forces alone are determinative. The big bang theory postulates an explosion at a zero-point 13.7 billion years ago, an inflationary universe rapidly accelerating and expanding. This theory is based on a number of assumptions that seem to fit the data, including the assumption that light travels 186,000 miles per second.

Astronomers use the Doppler effect to determine the velocities and distances of nearby stars. It is also used to ascertain that the galaxies are receding from us at rapid velocities. It is based in large part on the analysis of the light from stars and galaxies in outer space that reaches Earth and is photographed by our telescopes. The Doppler effect was introduced by Christian Doppler, a professor of mathematics in Prague. It was used to analyze both sound waves and light waves. The light from the stars moving away from us would have longer wavelengths, and the light from these would be shifted to the red on a spectroscopic analysis. Light from stars moving closer to us would have shorter wavelengths and would be blue.

The classical idea of a changeless universe is based on earlier science. At the end of the nineteenth century, scientists thought that our universe was "static" and "eternal." They were impressed by Newton's theory of gravity, which was able to predict the motion of objects in our solar system with precision. This cosmic perspective has been rapidly modified in contemporary astronomy by dynamic theories of the solar system and universe. The *dynamicists* point to catastrophic events in history, which have altered planetary, solar, and galactic configurations and led to radically new interpretations in astronomy.

Today we find that the universe is often in a state of violent turbulence. Hubble discovered that galaxies are enormous islands of stars. Astronomers recognize that these are expanding outward at velocities approaching the speed of light. We can go backward in time because the galaxies that we see today in our telescopes show events that occurred millions and billions of years ago. We can infer retrospective scenarios to the first explosion of space, which sent matter hurling at terrific speeds—the big bang. This infers that all the galaxies at one time were much closer together; hence, the concept of the big bang and the inflationary universe. This is known as the "standard model," which is beautifully depicted by Nobel laureate Steven Weinberg in his book *The First Three Minutes*.[5] He quotes the English historian of science Agnes Mary Clerke, who, writing in 1893, said that the immensity of the universe has been on a scale of magnitude such as the imagination recoils from contemplating—how much more so at the end of the twentieth century!

The apparent glow of background microwave radiation everywhere is said to confirm the big bang. But this tells us nothing about what happened before the zero-point expansion, how it occurred, or what kind of physics prevailed before that time. Nor does it exclude the possibility that separate isolated systems of causality and chance exist on planets and solar systems independent of the vast cosmic forces at work in separate island universes and galaxies such as our own.

The first astronomers of modern science presupposed the Copernican-Galilean-Newtonian heliocentric theory. According to the cosmological argument, God was the First Cause, an unknowable cause. Who or what caused God is allegedly unknowable. But nonetheless it presupposed a conception of the universe that was purely *materialistic* (there was no account of, or allowance for, the evolution of life), *mechanistic* (in terms of a machine model, not functional explanations in biology), and *deterministic* (in which fixed materialistic-mechanical causes explained the physicalist universe, leaving out entirely intentional explanations).

In the twentieth century the theories of relativity and quantum mechanics were both introduced, each questioning the notion of deterministic causality. How do these developments in the twentieth century fit within classical physics? There is no unified theory as yet that accomplishes it. In addition, the sheer dimensions of the universe (especially since the Hubblean revolution) have extrapolated our conception of the scale of the universe on both the macroscopic and microscopic levels. Moreover, the acceleration of the universe has dwarfed our comprehension of the universe. Can we simply assert that a unified field theory applies to *everything* in the expanding universe, a presumption for which there is presently insufficient evidence? Perhaps it is the physicist-astronomer who has engaged in speculative cosmological extrapolation.

Do the new cosmologies make due allowance for life in the universe? The quest for extraterrestrial forms of life is one of the most adventurous endeavors of human exploration, and although there is insufficient evidence as yet that our planet has been visited by extraterrestrial flying objects, in no sense do we wish to exclude organic matter from other galaxies or solar systems in outer space. But I think that present-day astronomy has not as yet accounted for extraterrestrial or extragalactic life. To do so may be a form of downward causality; namely, that it will need to develop categories and theories to deal with organic systems in outer space.

It is one thing to recommend that the basic natural sciences of physics and chemistry should be a major form of inquiry (of which I am in full accord), but it is quite another to translate that agenda into a cosmic principle. I believe that this form of the physicalist agenda should be interpreted primarily as a *normative methodological recommendation*; namely, that we should seek basic physicalist explanations on both the macro- *and* microlevels. On the macrolevel, classical mechanics and relativity theory apply; on the microlevel, quantum mechanics introduces probabilistic causality and chance.

The physicalist agenda is best interpreted as a strategy of research; it states that wherever we can, we should seek the most basic physical-chemical explanations possible. What we should *not* do is exclude explanations from other sciences that may be relevant but that are not physicalist in a strict sense.

Many researchers in neuroscience believe that explanations of consciousness ("mind" in earlier terms and/or "psychological behavior" in more recent terms) should be grounded in only physicalist explanations; that is, mapping the neural-geography of the brain and its physical-chemical structures in order to explain its functions. What about supervenient qualities? Do they not have some ontological status? Will the science of neurological networks in the human brain suffice to explain all forms of human behavior? It is presumptive to confidently assert that at present.

THE ATOMIC HYPOTHESIS

If we approach these questions not from philosophy but from the physical sciences, the role of the atomic theory in modern science causality is the basis of our explanations in the physical world of how things happen the way they do. The inductive-deductive mode of explanation serves as a successful model of what we want to accomplish in research. From the standpoint of physics, the basic ideas were derived from Newtonian science. The laws of mechanics enabled physicists and astronomers to account for the movement of the known planets around the Sun. By using Tycho Brahe's meticulous observations, Kepler was able to plot the orbits of these planets. The observations were of a fairly stable solar system in which regularity was the norm; hence it was thought that a deductive system applied in which phenomena could be predicted with accuracy. The physical changes observed operated in terms of deterministic forces.

In the twentieth century Einstein's special and general theories of relativity were introduced to relativize space and time and to modify classical physics. The atomic theory also became prominent. According to Richard Feynman, Nobel Prize recipient and popular Caltech physicist, the atomic theory became the cornerstone of the microscopic world of observation. He maintained that all things are composed of atoms and that we can understand something in terms of its constituent parts.

Feynman stated, "I believe it is the atomic hypothesis, or the atomic fact . . . that all things are made of atoms. Little particles that move around in perpetual motion, attracting each other when a little distance apart but repelling when being squeezed into one another."[6] The atom is itself divisible into subatomic particles. At the center is the nucleus, composed of protons and neutrons, which are encircled by electrons, with a large amount of space in between. The electron (negatively charged) is attracted to the proton (positively charged). Here a serious complication emerged. Werner Heisenberg introduced the "uncertainty principle," stating that it was not possible to determine the position and velocity of the electron simultaneously, as the electron takes quantum leaps from one orbit (shell) to the next.[7] This appears to inject a mode of chance into the most fundamental principles of nature. Eventually still other subatomic particles were postulated, such as quarks, positrons, and neutrinos.

This seems to undermine deterministic causality in nature. Was this due to our own ignorance, since we are unable to determine the position and velocity of the electron at the same time, or *is uncertainty inherent in nature as a generic trait*? According to quantum mechanics, we can ascertain statistical accounts, but this is probabilistic, not definitive causal knowledge. Einstein rejected this implication and attempted to work out a unified field theory that would include both relativity physics and quantum mechanics, without an opening to "chance." Speaking parsimoniously without any reference to the rule of chance in the universe, quantum physics can be interpreted as a mathematical machine that has been used effectively in applied fields to predict the behaviors of microscopic particles. Using statistical probabilities, it is the mathematical instrument by which we can explore their behaviors.

This is only a brief thumbnail sketch of some basic concepts in contemporary physics. The key questions to consider are (a) whether physics provides basic principles that govern *all* things, events, and processes in nature in every layer of the onion that we peel, and (b) whether other sciences, such as chemistry, microbiology, psychology, and the social sciences, can be deduced from the basic laws of physics. If this were the case, this would mean that physics indeed would be the foundational science from which all other sciences are derived.

PHYSICALIST REDUCTIONISM

Can the goal of reductionism serve us? Is reductionism to be construed as a basic theory of nature in the sense that nature is *nothing but* the world of atomic (or subatomic) particles, that is, the principles and laws governing mass and energy? The physicalist thesis states that reality *at root* is basically physical (or physical-chemical). All other entities, such as gods, spirits, souls, minds, and consciousness, do not exist as independent entities. With this I concur. In one sense physicalism is traditional materialism.

However, one could say in response that there is insufficient evidence for the total physicalist thesis. One may ask: Is the above generalization a scientific or philosophical claim? It is surely not "scientific" in a full sense, because it has not been verified experimentally; indeed, one needs to ask: How would one go about doing so? The sweeping physicalist theory is not like the theory of gravitation, for example, which is well grounded experimentally. I submit that reductive naturalism (it is "natural" because we can find no evidence for occult causes or entities) basically expresses a kind of pious hope, a leap beyond the evidence. If so, it can become a form of *seductive* naturalism, not totally dissimilar to *seductive supernaturalism*. It may even be a form of religious dogma or *belief*, because it is surely not a confirmed principle or generalization in the same sense as the theories of gravitation or natural selection.

Now let me hasten to qualify my criticisms by again maintaining that *reductive physicalism* does make sense, but *only* if we interpret it in strategic *methodological* terms, primarily as a program of research; namely, that we ought always to seek to find as far as we can the simpler and yet more basic explanations of phenomena by discovering more fundamental and comprehensive microexplanations.

Physics has made impressive progress in developing basic laws that explain the behavior of matter everywhere in nature. There is a kind of elegance in its view of how matter in the universe operates. The application of mathematics has enabled physicists to formulate, measure, and predict with accuracy these phenomena.

Physicists have discovered four forces in nature: first is the force of gravitation, as a universal principle, for mass and energy attract and are attracted to other matter. Newton showed that gravity is universal: it applies to the behavior of bodies on Earth, to the heavenly bodies in our solar system, and on the microlevel.

The second force is electromagnetism, which is produced by electrical charges. Electrons within atoms are electrically charged, as are the nuclei of atoms: like electrical charges repel, and unlike charges attract. In the nineteenth century James Clerk Maxwell developed the theory of the electromagnetic field, and he applied a series of equations to quantify the interactions of electrical and magnetic phenomena. Both gravitation and electromagnetic forces operate long-range. Gravitation applies to objects on Earth but also to large-scale phenomena, such as galaxies, black holes, and the expansion of the universe.

Third, there is also a strong force that acts between subatomic particles. This force binds quarks together in clusters to make subatomic particles, neutrons, and protons. This force also holds together the nucleus of the atom and undergirds the interactions between quarks and other particles.

The fourth force is the weak nuclear force that is responsible for some nuclear phenomena. The most familiar is beta decay and associated radioactivity.

Murray Gell-Mann, the codiscoverer of the quark, finds it remarkable that nature is "consonant" and "conformable" to itself and that these theories are widely applicable throughout the universe.[8] Like the skins of an onion, as we peel them, we penetrate deeper into the levels of the structure of the elementary particle system. Nature, he says, exhibits "similarities between one level and the next." Thus "the fundamental laws of nature describe the successive layers." It is likely he believes that it ends up as "a unified theory of all of the forces of nature." It would be a "unified quantum field theory of the elementary particles and interactions."[9] This, if achieved, would unify Einsteinian general relativity theory of gravitation and quantum mechanics. Such a simple unitary theory would account for all elementary particles and forces of nature.

However, Gell-Mann adds that nature involves more than physics and chemistry. We need to make allowance for chance events in the history of any heavenly body. In regard to Earth, it would include natural selection and evolution. He thinks that superstring theory may provide such a unified theory. This theory predicts an infinite number of forces, but most of these are too short-range to detect, at least at present.

Gell-Mann responds to the question of whether life on Earth involves more than physics and chemistry, and he answers yes, the *emergence* of new levels. "Life," he says, "can perfectly well emerge from the laws of physics plus accidents, and mind, from neurobiology."[10] He observes that although the "reduction" of one level of organization to a previous one is possible in principle, it is not by itself an adequate strategy for understanding nature. New laws and new phenomena appear and need to be accounted for in their own terms, to which I respond *hosanna*!

Steven Weinberg disagrees with the emergent theory and is a strong defender of reductionism, that is, he considers particle physics to be basic to contemporary physics, not merely a strategy of research. In his book *Dreams of a Final Theory: The Scientist's Search for the Ultimate Law of Nature*,[11] he offers two cheers for reductionism, quoting from the song "You and I," in which the singer croons to her lover, "you and I know" why the sky is blue and the birds sing melodies from the trees.[12]

Thus, he thinks that reductionism is a statement of the order of nature that is simply true. He grants that different levels of experience call for description and analysis in different terms, although he says that "we can never eliminate the accidental and historical elements of science, like biology and astronomy and geology." The hope is to trace the explanation of all natural phenomena to final laws and historical accidents.[13]

I reiterate that we should strive for a final theory (as a research agenda), but this, I submit, is within the context of physics and chemistry, and I do not believe that it is possible or even desirable, at this stage at least, to deduce explanations in all fields from "the final theory." Although I think that Weinberg has made important contributions to physics, his insistence upon the final theory being rooted in physics alone seems to me to be a form of special pleading, for there are good reasons to think that such a theory may not be able to subsume all other theories under its rubric.

The best illustration of the limits of reductionism, I submit, is in biology. Noted entomologist and biologist E. O. Wilson maintains that there are two fundamental laws of biology: "The first is that all the known properties of life are obedient to the laws of physics and chemistry." I agree that the laws of physics and chemistry are necessary conditions of understanding biology; the question is, are they sufficient? Wilson maintains that there is a second fundamental law of biology: that "all biological processes and all the differences that distinguish species have evolved by natural selection." He illustrates this by reference to the division of cells, which, he says, "are emergents." "They arise from the interactions of the molecules."[14] And, he says, "the movements cannot be readily deduced from principles of physics and chemistry." An emergent property is "so complex and poorly understood" that it must be dealt with by a different vocabulary.[15] He states that most of biology concerns emergent properties, which, for the time being at least, are only loosely connected to the laws of physics and chemistry.

The closest relation between biology and the physical sciences is DNA, the molecule that encodes heredity. This has given birth to the fields of molecular biology (which deals with molecules) and cellular biology (which deals with the processes and interacting elements of the cell). Beyond molecular and cellular biology is the rest of biological research. This includes the principles of natural selection, where we have seen that many macrofactors are included: differential reproduction, mutations, adaptations, the struggles for a survival of species, and more. Evolution involves change of a species through time, and this involves meticulously discovering and arranging fossils and bones in sequences of descent. Paleontology is thus related to evolutionary biology. Although genetic mutations occur, understanding these is insufficient for understanding the competition between males in an effort to mate with females of a species, or the selective role of females in choosing mates, and these factors surely do not reduce to microevents on the level of particles. Nor do physics and chemistry help scientists to understand the decline of biodiversity and the extinction of species.

Over and beyond this is the examination within organismic biology of the functions and networks of the organs of the body on the level of homeostasis, and similarly, in dealing with the invasion of the body by pathogens, which must be combated by white blood cells, or of threats from the external world. Nor do particle physics or chemistry enable us to understand how ecosystems exist in nature. Thus there are higher-order qualities that emerge—psychological phenomena, such as "consciousness" and "mind," for example—which cannot *ipso facto* easily be reduced to neurological states of energy, though they do not exist separate and distinct from them. There are still other newer fields of medicine that have emerged, such as epidemiology and immunology. Whether these in toto can be deduced from the laws of physics and chemistry still remains to be seen. Thus we may conclude that physicalist reductionism is mistaken, and that some form of nonreductive naturalism seems a more appropriate account of the contingent-random universe in which we live.

A PLACE OF DEATH—ASTRONOMY

The classical view that the universe is a serene place, hospitable to human life, is thus sheer fantasy. Moreover, to claim that everything has been "fine-tuned" by some intelligent Creator is anthropocentric wishful thinking gone wild. It depends on what is meant by the "universe." It is surely not true of our planet, which is the scene of violent storms that have the fury to destroy habitats, nor of volcanoes, which spew noxious chemicals into the atmosphere, nor of earthquakes, which unannounced can wreck vast sections of habitation. Nor is it true of ice ages, which have engulfed the planet, later followed in turn by global warming. Nor is it true of our solar system. As far as we can tell, our planet is the only one that contains life; though other planets, such as Mars, may have supported various forms of organic life in the past, and may still, though on a simplified scale. Either the other planets are too hot (such as Venus and Jupiter) or too distant from the Sun and too frigid (Uranus and Neptune) to sustain life as we know it.

The question is whether our solar system is a place of tranquility, in which the observations of scientists indicate perfect harmony and order, or a place of turbulence and disorder. After all, the elliptical orbits of the planets can be charted, and we can calculate detailed information about them and predict with accuracy where the planet will be next year or fifty years from now. However, one should not overlook the fact that our solar system is a place of collision and death. We have already alluded to the hypothesis that the dinosaurs became extinct some sixty-five million years ago due to a violent impact with a large asteroid, which caused the death of plant life and, as a result, their extinction. That hypothesis is speculative, but it is not without some historical evidence to support it. Reports of meteor showers go back through the millennia, as seen by the trails of light that follow them as they enter Earth's atmosphere.

The best case of a major impact of a foreign body is fairly contemporary, occurring on June 30, 1908, in a remote area of Siberia, close to the Tunguska River. The asteroid, which was perhaps seventy meters long, shattered as it entered the Earth's atmospheric envelope with an enormous explosion of energy. According to astronomer Philip Plait, this was hundreds of times the amount of energy released by the atomic bomb dropped on Hiroshima.[16] The explosion was viewed by many witnesses in Russia. The damage to the trees rippled out hundreds of square miles. No crater has been found, since the asteroid was vaporized several kilometers above the ground. A meteor crater can be viewed in Arizona, where an asteroid impacted the Earth some fifty thousand years ago.

Actually, our solar system is predominantly hostile to life, surely in the stratosphere and outer space, which are too cold and do not contain a conducive atmosphere: oxygen for organisms and carbon dioxide for plant life. In reality, our solar system is littered with debris, with belts of rock and refuse that sometimes collide with other bodies. Please note the pockmarked face of the Moon, caused by earlier volcanic eruptions and by falling meteors and asteroids. The same thing is true of Earth and other planets.

Plait estimates that the Earth is pummeled by twenty to forty tons of meteors every day.[17] Most of these burn up as they enter our atmosphere, but many of them crash, some of them with disastrous effects. The Earth is constantly plowing into debris it encounters in space, which is swarming with junk. Most of this debris is made up of small meteorites, but some are fairly large asteroids, which have impacted Earth with devastating results. Most of the rocks come from an asteroid belt between the orbits of Jupiter and Mars, and there are estimated to be billions of them. Most of them are very small, but some are fairly large. Comets have also been observed in our solar system, such as Halley's comet, whose ellipse has been charted. Comets are made of gravel and rocks and held together by ice. These can come from anywhere in outer space. The Shoemaker-Levy comet eventually slammed into Jupiter and was destroyed on impact. The Sun is not a silent body. Its surface manifests "sun spots," which are magnetic storms that spew enormous amounts of energy throughout the solar system.

WHAT ABOUT OUR EARTH?

Humanity has always speculated about the origin of mother Earth, upon which we dwell. Creation myths have been laden with poetic mythological symbolism; often such revelations were allegedly delivered from On High, on a mountaintop by religious prophets who declared that God created the heavens and the Earth. Moses encountered a burning bush, from which Yahweh spoke to him, offering directions of what he was to do, and later the Ten Commandments were allegedly delivered to him on Mount Sinai. Mohammed had many encounters with the angel Gabriel as he wandered the caves and mountains outside of Mecca, from which the Koran was first revealed. Noah's Ark, after the Great Flood, was supposedly shipwrecked on a high mountain in Turkey; and according to the Homeric myths, the mighty gods of Greece occupied Mount Olympus. In our own age we have no shepherd's rod and staff to comfort us; instead, we have the tools of telescopes and human-made satellites, which soar outward and above.

Planet Earth, as seen and photographed from outer space, according to Carl Sagan, is a "pale blue dot." At one time it was thought to be a perfect sphere spinning on its axis every twenty-four hours, orbiting around the Sun every 365 days. Our new perspective from outer space views it as we come closer as green overgrowth of plants and trees, with continents surrounded by blue oceans and seas; and we can even see the white polar caps. It is a tranquil picture of idyllic beauty.

Astronomers have attempted to speculate about its probable origin. It was not created by divine fiat in six days, as Genesis declares; rather it is the result of a turbulent past, a product of cataclysmic events that astronomers and geologists have attempted to reconstruct, using the best tools of modern science. They have come up with possible scenarios: the big bang scenario estimates the origin of our universe as 13.7 billion years ago, a violent episode with an inflationary universe and galaxies that are rushing outward at enormous speed. The age of the Earth has been estimated at approximately 4.6 billion years. The discovery of radioactivity enables us to place a timescale by dating rocks that we uncover on our planet. These rocks often contain radioactive elements that decay over long periods of time.

Earth's origin is closely related to the creation of our Sun and solar system. The most current hypothesis is that our solar system was a rotating gas cloud and interstellar dust that may have been perturbed by a passing nearby celestial body, or perhaps even a supernova. The dust and gas began to collapse inward because of gravitational attraction.

Earth's surface was molten at its inception. As it cooled, volcanoes erupted, spewing forth great amounts of carbon dioxide. The oldest volcanic rock has been dated as approximately 3.75 billion years old. Meteorites created at the same time as planet Earth indicate that they are probably 4.5 billion years old.

In reconstructing a historical sequence, the first evidence of bacteria appeared 3.8 billion years ago. According to this account, life most likely began on Earth some 700 million years after the planet was formed, existing in shallow oceans near thermal vents and minerals. Other elementary forms of life appeared when plants became able to convert carbon dioxide and water into food, using energy from sunlight by a process called *photosynthesis*, producing oxygen as a by-product, which was emitted into the atmosphere. Most of the carbon dioxide became locked into sedimentary rocks as fossil fuels and carbon atoms. As oxygen in the atmosphere built up, an ozone layer was formed, which filtered out ultraviolet rays and allowed the emergence of a wide variety of living organisms.

This is a brief sketch of the origins of our planet, based on circumstantial evidence and conjecture. Ever since that time, other forces have stamped new prints on our Earth, which is composed of a hot nickel and iron core and an outer mantle primarily of silicates with a thin covering surface. Earth underwent orbital variations, cosmic bombardments, the shifting of its magnetic field, volcanoes, earthquakes, and the movement of tectonic plates. The findings of German geologist Alfred Wegener were of considerable influence. In 1912 he said that there was evidence of continental drift and movements. There are ongoing processes of erosion, rivers carrying mud and sand to the deltas and seas, the chemical breakdown of rocks such as limestone and sandstone, weathering, periods of cooling down and the growth of glaciers and of warming up, and of course the appearance of various life forms. The impact of human civilization on the planet has been particularly devastating; humans today dominate the planet. It may be even considered as a living planet, because the ecosystems of the biosphere are interdependent. Thus Earth has been undergoing substantial changes throughout its history.

Another hypothesis is that the Earth (and the Moon) may have been formed out of rocky snowballs, which probably grew hot. Another theory is that a large celestial object grazed the Earth some four billion years ago, causing a shower of rocks to break loose, which collected together to form the Moon. Its surface shows a substantial number of craters caused by asteroids and meteors.

HOW DID OUR SOLAR SYSTEM EMERGE?

The most likely explanation of the formation of the solar system is now supported by observations from the Hubble Telescope. The evidence supports the hypothesis that it was at first a protosolar nebula, resulting from "the gravitational collapse of a cloud of interstellar dust and molecular gas." Planets eventually formed out of rings of dense concentrations of dust and gas, including Earth. These eventually became the inner planets of the solar system.[18]

The planets of our own solar system likewise show a violent past of volcanic activity, similar to Earth and its Moon. For example, circumstantial evidence has been used to ascertain how the surface of Mercury was formed. A NASA spacecraft flyby of Mercury transmitted a great deal of information, including the confirmation that volcanism occurred on its surface. Photographs of Mercury's surface indicate that volcanoes played a role in causing plains on the planet. The smooth plains point to volcanic deposits similar to basaltic maria (seas) on our Moon.

Similar observations and conjectures have been used to explain the origins of other terrestrial planets. The volcanoes that ravaged the planet Earth, and still do, occurred on Mars as well. Similar findings indicate volcanic activity on Venus. Scientists working at the European Space Agency's *Venus Express* spacecraft mission point to the similarity with Earth, though the planets are different. Venus has approximately the same composition, size, and mass. Its surface temperatures, however, are much hotter than those of Earth, at over 800 degrees Fahrenheit. This is most likely due to the heat-trapping effects of its atmosphere, which is almost pure carbon dioxide and much denser than Earth's surface. Such studies draw on analogous causal hypotheses—the surfaces of planets and satellites in our solar system are pockmarked by the impact of asteroids and volcanic activity past and present. Yet since different asteroids hit different planets at varying velocities, these planets have different histories, and their surfaces and atmospheres (if they have any) thus may differ. Similarly for the various moons that encircle Jupiter and Saturn.[19]

The planets bear similar characteristics; they are basically circular, spin on their axes (at various speeds), and orbit the Sun. Venus and Earth appear roughly similar; some consider them to be twins because Venus is approximately the same size and has a similar mass and composition. Venus differs in that it is closer to the Sun, with the light shed by the Sun twice as intense as Earth's and hot enough to melt metals such as lead and tin. Venus has a much thicker atmosphere, however, which is composed of carbon dioxide and clouds of sulfuric acid. This is a hundredfold more dense than Earth's.

These observations have been obtained from the European *Venus Express*, the space satellite that probed the planet. The *Mariner 2* probe reached Venus in 1962. Russia has also successfully sent probes to Venus. Scientists infer that Venus was formed with large quantities of water, but because of the Sun's heat, it evaporated and was transformed into water vapor clouds, a greenhouse gas that traps the heat in the atmosphere. All of the planets in our solar system have been visited by robotic spacecraft equipped with close-up cameras and the means to conduct tests on their atmospheres and surfaces.

Similar questions have been raised about Mars, onto which human spacecraft have been able to land. There is some evidence of the presence of water and its erosion of the surface of Mars in the distant past. These planets are similar yet idiosyncratic bodies that differ from one another. Our solar system is composed of eight planets—Mercury, Venus, Earth, Mars, Jupiter, Saturn, Uranus, and Neptune—all with their own unique histories. It has four dwarf planets: Haumea, Pluto, Makemake, and Eris, which orbit the Sun beyond Neptune and Ceres in the asteroid belt. A planet is any body in orbit around the Sun that has enough mass to form itself into a spherical shape and has cleared its immediate neighborhood of all smaller objects. Each planet, however, is different in makeup, heat, and surface. There are 166 known moons orbiting the eight planets in our solar system. There are in addition billions of small bodies, including asteroids, the icy Kuiper belt, comets, meteorites, and interplanetary dust.

SHIFTING ORBITS

As we saw, Newton's model of the solar system was analogous to a clock that kept exact time. The orbits were precisely fixed through time. The alternative cosmic perspective is that of a *dynamic* universe that has violently changed throughout time, and there is nothing so supportive of this vision as the empirical evidence that the orbits of the planets in our solar system reshuffled during the early years of the solar system. This points to a cataclysmic solar scenario. The "Nice model," as it is called, was first proposed about a decade ago by astronomers from the observatory of the Côte d'Azur in Nice, France (not far from where I spend part of the year).

The great reshuffling was a short and violent affair that placed the outer planets (Saturn, Uranus, and Neptune) where they now are, created the Kuiper belt beyond Neptune, positioned oddly orbiting moons around the outer planets, and rained bombardments of asteroids and comets.[20] Why the orbits of the inner planets were not affected is a puzzle. Nonetheless, its proponents maintain that it best accounts for the data. If so, it is further evidence that the dynamic model best explains many historical changes on the astronomical level.

HOW WAS THE SUN FORMED?

The Sun was born in the later stages of the universe some 4.6 billion years ago. The Sun contains 99.86 percent of the solar system's known mass. It has a large gaseous interior density, hot enough to sustain nuclear fission, which releases energy mostly radiated into space as electromagnetic radiation or visible light. Hydrogen nuclear fusion is generally hotter and brighter. It also emits a continuous stream of charged particles (plasma) known as the solar wind. There are geomagnetic storms on the Sun's surface, solar flares, and coronal mass ejections.

Our Sun is part of a galaxy of suns or stars. Galaxies have spiral arms, which are composed of stars, dust, and gas. Our galaxy, the Milky Way, is a spiral galaxy one hundred thousand light-years across with four hundred billion stars.

There are several theories about the origins of the solar system. One scenario is as follows:

- It was formed from a solar nebula, an immense rotating cloud of dust and gas.
- A nuclear reaction began at the dense center of the nebula.
- The planets were formed as a result of material accumulating within the gaseous cloud.
- Planets near the Sun evolved as relatively small spheres of rocky material.
- Further out from the solar nebulae, gasses and debris accumulated to form large gaseous planets.

The cloud of interstellar gas and dust may have been dislodged by the explosion of a nearby star (a supernova). It collapsed into a center by gravity.

In less than one hundred thousand years, this became compressed and heated enough for the dust to vaporize. The center eventually became our Sun.

SPECULATIONS ABOUT OUR GALAXY

The Milky Way was probably born 12.3 billion years ago, 1.3 billion years after the big bang. It may have merged with its sister galaxies, acquiring dwarf galaxies and star clusters and merging with a sister galaxy. Our galaxy is expected to impact with the Andromeda Galaxy in the future, some three billion years from now.

Interestingly, our Sun orbits the center of the Milky Way once every 220 million years. The oldest stars have their velocities disturbed by passing close to heavy objects in the Milky Way, such as giant molecular clouds, supernovas, spiral arms, and by the capture of satellite galaxies.

The history of the Milky Way is thus turbulent. Stars are not fixed in space; they move around, sometimes rushing through other galaxies or approaching close enough to one another for brief cosmic trysts. Our Sun moves toward and then away from the center of our Milky Way and moves up and down through the galactic plane. One complete up-and-down cycle takes sixty-four million years, which, incidentally, is close to biodiversity and extinction cycles. The Milky Way is being pulled gravitationally toward a massive cluster of galaxies called the Virgo Cluster.

Cosmic rays from outer space influence life on our planet and can lead to mutations. Our solar system and galaxy travel through outer space at great speeds, hundreds of thousands of miles per hour. They bond to other stars in the Milky Way by gravitational force. Gas clouds and various forms of radiation have constantly showered our Sun and planets, including planet Earth.

Is the universe open or closed? Will it continue to recede at increasingly rapid velocities, or will it someday snap back like an elastic band? This is difficult to say for sure, because we know only a small part of the total picture. The space age has enabled the human species to peer further than earlier generations into outer space; the second half of the twentieth century was an age of space travel and of space satellites that enabled our imagination to soar even further. Our observations point to cosmic turbulence on a vast intergalactic scale.

Two things are especially intriguing: first, the discovery of supernovae, and second, the discovery of dark matter.

The Hubble Telescope and space satellites have transmitted dramatic images of colliding galaxies deep in space. The *Kepler* telescope and others in the future will extend our reach into the unknown, even farther than today. They show galaxies spinning, sliding, and slipping into one another. This stellar destruction apparently gives birth to new and larger galaxies. Out of these collisions remarkable forms of intricate structures are seen. These processes take hundreds of millions of years for galaxies to merge, and the light from these distant collisions takes hundreds of millions of years to traverse space. During the supernova explosion of a star in outer space, its luminosity increases manyfold as the bulk of the star's mass is blown outward at a very high velocity, even the weakest supernova SN2008ba had a brightness of about twenty-five million suns.

The most recent known supernova explosion in our galaxy was in 1572. It was viewed by the astronomer Tycho Brahe, who was deeply impressed by its brightness, which rivaled that of Jupiter. This was the constellation Cassiopeia, which is located in the northern sky on the edge of the Milky Way; it is a constellation of five stars, named for the mythological Ethiopian queen Cassiopeia. Supernovae are the violent deaths of stars. The hypothesis proposed by astronomers is that this massive star first ran out of fuel and collapsed. The consequent explosion ejected the outer layers, leaving a cloud of gas now seen in an image photographed by telescopes. What remains is a leftover core, which seems to be a neutron star that is immensely compacted.

Astrophysicists Lawrence Krauss and Robert Scherrer maintain that the rapid increase in the expansion of the universe will eventually hurl galaxies faster than the speed of light, which will lead to their dropping out of view.[21] It would erase any signs that the big bang occurred, and what will happen will be beyond the horizons. This means that the past will be lost at some time in the future, at least to observers. The universe was inhospitable to life when the big bang occurred—it was too hot—and it may become inhospitable if and when our solar system cools down at some remote time in the future.

The universe continues to reveal strange dimensions. What are these? Gamma-ray outbursts belch more energy in a second than has been produced by our Sun in the past one billion years; black holes absorb entire galaxies. The explosions of supernovae scatter a vast array of elements on stellar winds.

Our concepts of the universe are apt to change still more radically. We must guard against any anthropic temptation to read our own longings into nature, which is so vast that our puny existence is a mere flutter in a virtually infinite cosmic sea. There seems to be no discernible purpose to the universe:

1. Clearly massive physical and thermonuclear forces interact on the astronomical level. But contingency and chance also intervene—as seen in both quantum mechanics and galaxies. What is clear is that the scale of the universe is beyond any previous comprehension. Some recent estimates have put it at 10 to the 500th power, or 10^{500}. The size of our universe could be infinite, but observations say it is at least ten times bigger than we can now observe.
2. Turning back to Earth, geology shows the results of erosion, collision, solar bursts, and climatic changes. The biosphere indicates that cosmic bombardments are contributing causes of the contingencies and harmony, order and chance that we observe. The human world of individuals and societies, history and individuation likewise manifests the same processes of turbulence and chance, and the efforts by humans to create order and harmony.
3. There are in the universe at large general laws and unique events, historical individuation and general coduction.

Astronomers discovered only in recent decades that the largest part of our universe had been hidden or unknown to humankind. Some consider this discovery of dark matter the most important in the past half-century. Astronomers recognized that the universe was expanding at an increasingly rapid rate of speed. It was assumed that this expansion would slow down at some time in the future, for they believed that gravitational forces would eventually reduce it and perhaps bring it back. They postulated hidden energy that opposed gravity and forced the galaxies to move farther apart at a faster rate. This is called dark energy because we do not see it, though most of the universe is composed of it. If dark energy intensifies, this may in time yank apart the galaxies, stars, and planets, causing them all to gradually cool down in the vastness of space. If dark energy weakens, gravity can again exert its force, which would pull the parts of the universe back into a "big crunch."

Much of this is fed by mathematical extrapolations from the analysis of light, with the shift toward red indicating a receding universe, expanding outward. Perhaps the most that we can say at present is that the cosmic picture now being presented is still shrouded in mystery. One thing is clear, however: it is a strange, restless, changing, turbulent universe. I think that the term *pluriverse* perhaps describes better what we have encountered. We need to be skeptical of all that we have not been able to confirm experimentally. We simply need to assume the posture of the agnostic—not that of the supine and trembling believer, who bows before the magnificence and mystery of nature in prostrate reverence and obedience to vast unknown divine forces in the hope that some Being or beings out there will respond to his plaintive pleas for salvation.

My response as a critical thinker is to be impressed by what humans have been able to find out about nature, but also to recognize that our brain cells may not be able to fathom what is happening beyond the dim whispers of light that may come our way out of the deepest recesses of the pluriverse. Several things are by now clear. What we have observed and noted has enabled us to surmise many things about the vast worlds of nature. We find order and regularity, symmetry and balance, bonding and harmony. But we also find that there are many gaps in our understanding.

The universe is indifferent to our hopes and fears, uncaring about how things affect creatures on one planet on the edge of a galaxy. In the universe at large, many things have been noted. Things may go smoothly at times, yet there are often rough edges in the cosmic scheme: There is surely disorder and irregularity, conflict and violence in every sphere of nature. There is attraction and repulsion, disequilibrium and decay, growth and senescence. The implication is that we need to express an intense passion for life, because life is the ultimate affirmation within nature. Yet life is fragile and does not last forever. Life, as we experience it, contrasts ecstasy and agony, victory and defeat, success and failure; it is brittle and will someday be shattered.

Nature proliferates in fecund abundance, but it also destroys. The life world is replete with predators who consume other forms of life and savor its delights.

Humans raise and slaughter billions of gentle fowl, lamb, cows, and silent creatures of the deep—with little concern for which animal or fish they trap and fry in order to satisfy their hunger or delight. Nature is sometimes overcome by droughts and floods, cyclones and tornadoes, raging forest fires, and dazzling meteor showers. My prostate doesn't work well, cancer may grow wildly, or a supernova may burst, destroying everything within its perimeter. So it is often a dissonant life in a seemingly random universe that we experience. Whether that which exists does so in a totally interconnected *uni*-verse, *multi*-verse, or *pluri*-verse is difficult to fathom. From one vantage point it is both a *uni*-verse and a *dis*-verse.

Leading astrophysicists now occupy the throne of respect, especially if they are adorned with Nobel Prizes conferred on them upon recommendations of their colleagues who share the same mystique of scientific extrapolations. This place of honor was formerly held by discredited theologians.

The point is that some of the scientific hypotheses they develop can be applied as tools that we can use with effectiveness. They seem to work in enabling us to build bridges and skyscrapers, bore tunnels, shoot satellites into space, cure disease, reduce suffering, extend and enhance life. The proof of science is in consuming its pudding; but there are various recipes, some that work and some that do not.

Cosmological theories are like the reading of inkblots in a Rorschach test: How we interpret a blot says more about us than about the blot. Remember, the Ptolemaic epicycles enabled humans to calculate the paths that the heavenly bodies took, even though they were false. Thus various cosmological theories may be validated by the theories we concoct, without being true. This may be the secret to the big bang's postulation of a singularity and an inflationary universe, expanding at increasing velocities. And it may be true of string theory, which has a kind of mathematical elegance, though it may not be the ultimate account of the entire cosmic scene.

One inference that humans today may draw in viewing the planets, stars, galaxies, and clusters in outer space is the mind-boggling immensity of the multiverse. Humans have always been awed by the mystery of the night sky. Depending only on their unaided vision, primitive men and women were likewise overwhelmed by the mystery of the cosmic scene, and they propitiated hidden gods to assuage their fears and confer favors. Scientific culture has long since abandoned such anthropocentric accounts of nature. We are rightly impressed by the progress that science has made in interpreting nature. We have expanded our powers of observation exponentially by using modern technologies. Beginning with Galileo's simple telescope, astronomers today have magnified the range of their vision enormously with improved telescopes, space satellites, radio astronomy, and other instruments. They have drawn on mathematical theories to explain what we are able to observe, and to test these theories experimentally as best they could in the light of the data.

Yet over and beyond the achievements of the natural sciences and astronomy, there is still a deep sense of the unknown, and this can only lead to an increased feeling of *awe* and astonishment because the universe is boundless and we have glimpsed only a tiny part of it from our small planet encircling a minor star at the edge of the Milky Way. Some reverence on our planet for nature is fitting. It is both *human* and *natural*, and it should not be confused with the worship of mythical gods enshrined by the ancients. Some atheists may be incensed by these remarks, but if they are, they are insensitive to the magnificent splendor of the universe and our effort to probe its depths as best we can and to come to terms with its wonders.

Modern humans now have a new, open vista on the universe. Using the polished mirrors of telescopes, which serve as "spectacles," they are able to peer far into the beyond. Astronomers have applied the laws of physics and mechanics to calculate the gravitational forces of huge bodies from afar. They can send satellites into space to peer even further beyond Earth's atmosphere, and these have enabled humans to probe other distant solar systems. They have calculated that the stars and galaxies are receding at enormous speeds and that the dimensions of the universe are expanding outward.

What an incalculable leap of human imagination this will provoke. Scientific inquiry has transformed our planetariums, which chart our solar system, into *galaxceriums*, which display galaxies in outer space: our own Milky Way, Andromeda, Bodes Galaxy (discovered by Johann Bode in 1774), the Cartwheel Galaxy (which looks like the spoke of a cartwheel), the Cigar Galaxy (it appears similar to a cigar), the Large and Small Magellanic Galaxies (a large "cloud" named after Ferdinand Magellan), the Whirlpool Galaxy, and others.

Epicurus was prophetic when he declared that "there will be nothing to hinder an infinity of worlds." Our Milky Way has billions of stars, and many of these most likely have planets encircling them, some even with life. We have only discovered thus far approximately four hundred such planets because of the serious technical difficulties of locating them. How many others are able to sustain life because they are not too close to burn up or too far to freeze has yet to be determined.

French biologist Jacques Monod pointed out that life evolves through the working of natural laws, that is, natural selection, but contingency and chance also have a role. Thus the extinction of millions of species on the planet Earth left room for the emergence of new species. Undoubtedly, if life forms exist on other planets—and they probably do—then the kinds of life that will have evolved are indeterminate. In this sense the creativity exhibited by life on Earth most likely may be found in other forms of life on other planets in the universe at large.

ACT FIVE
THE EFFERVESCENCE OF EMERGENTS

NONREDUCTIVE NATURALISM

There is abundant evidence for the general statement that all entities and things, animals and plants, processes and events, fields and systems, are material at root. Although identifying basic physical causes is a necessary condition for understanding nature, it is not sufficient, because there are additional properties, qualities, and relationships that emerge from the physical base.

Minds would not exist without brains or bodies; they are functions of them. There is insufficient evidence that minds survive after the death of bodies. Works of literature or art were created by human beings, and the cultural artifacts and traditions that were developed and persist are often products of long historical processes and traditions. All of the above are nonreductive *natural* phenomena. There is a place for them in a naturalistic universe.

Change is a constant feature of existence. There are various sorts of processes that can be described in different contexts: There are rapid processes of change on the atomic level, and there are interactions between planets and moons, meteors and asteroids in our solar system. There are processes of erosion and bombardment, cracks and fissures, floods and storms on the geological scale. In living systems there are genetic and chromosomal interactions, division and reproduction, growth and development, senescence and death. In the biosphere there is natural selection, the evolution and extinction of species. On the galactic scale there is massive turbulence: the origin and eventual destruction of planets, moons, and meteors, and the birth and death of suns. Huge galaxies collide, supernovae explode, and new stars appear.

In recorded human history, change is ongoing. There are transformations and revolutions; wars and plagues; social and cultural changes constantly occur; there are military, political, and economic developments; intellectual, scientific, and technological discoveries—all of which continually remake the countryside, urban life, global trade, and communication.

There are so many things that we have discovered about the various aspects of nature. Planets and moons, solar systems and galaxies, volcanoes and hurricanes, forests and jungles, deserts and oases are all part of nature; so are swarms of bees, schools of fish, herds of cattle, fungi and insect colonies.

Nature also includes human beings and their villages, towns, and cities, countries and constitutions; and it includes cultural changes that are a result of human dreams and aspirations. There are literary and artistic trends: the literature of Shakespeare and Molière, the art of Miró and Brâncuşi. The human world includes Sappho and Hannibal, Epicurus and Montaigne, palaces and towers, books and testaments, moral beliefs and convictions, science and religion, poetry and opera. The world of nature and culture is chock-full of diversity and similarity, riches and poverty, disease and conflict, living organisms and extinct species, fossils and the remains of ancient civilizations that are no more, and splendid new ones that emerge in human history.

Nonreductive naturalism does not deny existence to anything or anyone. It includes whatever is, was, or will be. It contains corrupt ideas and malicious conduct: the good, the true, and the just, as well as the bad, the false, and the unjust. It includes ancient Rome and modern Siberia, the kingdom of the Incas and the British Empire, the Magna Carta and doctrines of human rights, communist revolutions and fascist oppression, theocracies and inquisitions, democracies and free societies, bridges and tunnels, pyramids and skyscrapers, the Taj Mahal and the Black Hole of Calcutta.

You will ask, is this full-blooded and lusty naturalism sufficient, or is it a dumping ground for whatever has been, exists now, will tomorrow or in future centuries? Are there no distinctions to be made? Are they all "real," asks the skeptic, the unnatural as well as the natural?

In one sense, yes, things exist as they exist under whatever sky or terrain, and in a multiplicity and diversity of contexts. This form of *metanature* does not banish from nature whatever is or was. To be is to be, no matter what, and in whatever guise it assumes.

Does nature include the illusions of humans, the misconceptions, and the speculative nonsense? By saying this, have I not trapped myself in a corner, forced to allow the Koran and the Hebrew Bible, the New Testament, the Upanishads, and the Book of Mormon on the same par as Euclidian geometry and $E=mc^2$, the theories of natural selection and gravitation, antibiotics, and cardiology?

No, obviously we need to distinguish what is *true* from what is *false*. If people believe in ghosts and poltergeists, the tooth fairy and the devil, they may exist in human imagination and culture at some time in history, and human beings act in the light of their beliefs. But they are *not* true accounts of reality, so we need some criterion of what is actually *real* about the world, independent of what people may be seduced into believing is true. We seek *reliable knowledge* about the world, unscreened by a cultural mask or bias. I am now sitting at my wooden desk, under a fluorescent bulb; I get up and walk downstairs to the refrigerator. This is real to me and to my pug and Siamese cat, as it is—in some sense—to the cockroach hidden from my view. Existence depends upon the context or situation or setting in which I transact. I can dream of the beach in Juan-les-Pins or the art museum in Saint-Paul-de-Vence, but I am not there now. So what is depends on sentient beings and their environments and exists in *that sense*. They cannot be exiled from the universe. Yet we have the *methods of scientific corroboration* continuous with common sense, and we can ascertain what is true *independently* of our subjective fancies or cultural mythologies.

Young children may be led to believe that Saint Nicholas bestows gifts upon them at Christmas and that an entire industry exists to create toys to whet their imaginations and sell products. They are part of a culture, but they are not true—there is no North Pole populated by little elves who manufacture presents and a giant set of reindeer and Santas to go down chimneys loaded with gifts for countless generations of children. Myths may exist in culture and be perpetuated, but they are not true of the world. Fairy tales are only *tales*, and these include the Jesus, Moses, and Mohammed myths, which promise salvation in paradise to believers and torture in hell to miscreants and backsliders. If culture is part of nature, how are we to distinguish between them? Are the myths of the

Seneca nation of Native Americans any less true than the wild stories brought to the colonies by believing Christians and Jews? Red Jacket was a well-known Seneca chieftain who fought with the British against the French during the so-called French and Indian War of the eighteenth century. He also was a party to the peace treaty between George Washington and the Seneca nation. In a classic speech he remonstrates against the effort of the Christians to convert the Senecas to Christianity, saying it was folly to do so because Christians raped Indian women, got their braves drunk, and were not of higher moral virtue than traditional native culture.

Obviously there needs to be some method of developing reliable knowledge about what is true or probable and what is false and mistaken—and that is the method of scientific corroboration, a pluralistic interpretation of scientific inquiry and validation. I should add that false faiths and mythical ideas have transformed the world: They have launched wars and destroyed peoples (such as during the Crusades, the jihads, and the Holocaust). And they have constructed great cathedrals, pyramids, temples, synagogues, and mosques to worship false deities.

But let me return to the basic question: Is reality reducible to its simplest components, as described by physics and chemistry, or are there emergent events, processes, properties, qualities, and relationships that coexist along with their rudimentary parts? A lovely illustration of this is gained by consulting my wife's French recipe book cherished in her memory as passed down to her from her mother's mother's mother—grandmothers galore—and what delicious menus she can prepare! Take crème caramel, for example, made of eggs and sugar and milk, caramel and vanilla. Blend them together, bake the batter in the oven, allow it to cool, and what a succulent dessert will emerge to smack your lips over and enjoy! Is the crème caramel the sum of its parts? Yes, of course—and the process of melding them, yes, *or* is it something more? The latter, I suggest, is true in one sense; it is what it is because of a creative process of whipping, baking, and bringing into being a new dessert that now exists; it is an *emergent* product, a result of a cultural tradition. Similarly for the wines and breads and cheeses that my wife purchases in the market where she shops while we are in France, and all the more so for the modern world of skyscrapers, automobiles, and satellites. The products manufactured by the technical and artistic skills of humans have become part of nature and are real, as are the nests built by birds to shelter and feed their young chicks, or the dams built by beavers. So the creative works of humans have changed the natural environment by felling trees or cutting stones out of quarries, or forging iron and steel, and they are *real* in the world. But what of misconceptions and fantasies? I have argued earlier that a reductive theory grounded on a basic physicalist theory of atomic and subatomic particles is insufficient to account for everything.

I have attempted to find synonyms for the general concept *emergents*. It refers to that which appears, surfaces, manifests itself, comes forth, evolves, grows, develops, or unfolds. Effervescence may involve turmoil and agitation. It also may entail the idea of opening up, flowering, blooming, or climaxing. In this sense it implies the *consummatory* aspects of existence, particularly in the life world. It is clear that emergents in some sense are *generic traits of nature*.

Intrinsic to the concept of emergence is *novelty*, uniqueness, newness. All of these terms recognize that to be is to exist not simply as a combination of material parts but also as an integrated functional whole, which begins to manifest qualities of *fluorescence* or *luminescence*, particularly as the emergence of novelty refers to *new* species or entities that evolve. This is by way of contrast with extinctions or deaths. So there are in nature both the *extinctions* of forms of life that have disappeared (and can be found only in fossil remains) and aspects of nature that display the *emergence* of evolvents—new, novel, unique forms that come into being as new entities. There are always new things under the sun, as it were, and these may refer in the physical universe to the emergence in time of new planets, or moons, or stars out of nova explosions, new galaxies or clusters of galaxies that merge and emerge.

The French term *naissance* (birth) is intrinsic to the concept of emergence, as something seen for the first time. Every infant is a unique emergent, as it were; similarly for every new species that gradually evolves from earlier forms. Similar phenomena are encountered in the world of culture. Every new application or invention is an emergent within human civilization, a creative accretion or addition to what has been. Often there is a *renaissance*, *risorgimento*, or radical awakening, and this may lead to new bursts of creativity, as new musical forms, schools of artistic expression. Political uprisings or revolutions may be violent and bring in the enunciation of new rights and freedoms for the dispossessed or disenfranchised. And out of this new charters and constitutions may be written and new social institutions may emerge. *Emergents* thus may apply to a wide range of phenomena that appear in a civilization and capture the commitment and devotion of a people, such as the new America, or the French Revolution.

The question has been raised about the status of emergent properties: Do emergent properties exist in nature quite independent of its parts or of human culture? This is especially pertinent to the biosphere where emergent properties are discovered. Here we learn that organic matter is able to ingest nourishment from its environment and to reproduce sexually. Is life a kind of emergent from inorganic matter? We see abundant evidence of emergent characteristics and of new species that come into being and older ones that became extinct.

The theory of emergent evolution postulates that at one time in the historical past, early groups of cells adapted to conditions in the environment, and new forms of behavior developed. Fish fins became amphibious feet as the fish learned to leave the water and creep onto the beach, adapting to life ashore. Others adapted, learning how to use skin flaps to fly, and they became either pterodactyls or bats. Many of these new characteristics evolved gradually by increments. Plasticity enables a member of a species to stretch and accommodate.

On other occasions a sudden chance mutation may occur, giving an individual who possesses the capacity an advantage in the struggle for survival: It is able to reproduce and transmit these new characteristics to its offspring, and in time it becomes more common and a new species may emerge, only able to mate with similar kinds. Stephen J. Gould and Niles Eldredge, in the paper "Punctuated Equilibria," describe a new emergent life form, as something like a gun cocked to go off, spring-loaded for change.[1]

What seems pervasive in the life world, as I have said, is the emergence of novelty. Interbreeding of isolated members of a species exacerbates the differences, as Darwin observed on his trip to the Galapagos; and we may still see today an analogous emergence of exotic forms of life in Madagascar or Australia.

Similarly for the appearance of a new planet, moon, star, or galaxy: each is unique, and what it is depends on its individual history, which has formed it, and in some sense it is an "emergent." The asteroid ring around Saturn is a good illustration of a unique, individual network of objects encircling a planet; even though other planets may have similar rings, they are unlikely to be exactly the same.

Could anyone—knowing the basic laws of physics—have predicted that such characteristics and combinations would emerge in the organic or inorganic world? I recall a memorable trip to Kruger Park, a wildlife preserve between the northern part of South Africa and Mozambique. As I was sitting quietly with my wife and father-in-law in a parked vehicle not far from a pond, two huge hippopotamuses came to drink and stare. One could imagine a conversation between them; one hippo saying to his partner, "Take a look at those three humans over there. Have you ever seen anything so ugly?" It is as if Rube Goldberg was in charge, concocting and creating the weirdest-looking creatures. Look at an ostrich, for example, with a pin-sized head, a long, thin neck, a big round torso, and plumed feathers thrown in. These are hardly a product of intelligent design but rather of natural selection, involving both lawlike regularities and random chance, emergent capacities for adaptation and survival. But the process of evolution is an established process with recognizable steps even though how those steps are manifested within given life forms may be quite different. So there is chance within order and order within chance.

There are new principles of organization that also seem to emerge. Is a whole the sum of its parts or more than them? The latter, defenders of holism have argued. We need not invoke separate holistic entities in nature, yet there are new qualities that are surely discernible on different levels of complexity. Some critics maintain that holistic qualities seem rather intangible, for they are difficult to verify as part of a theory. Yet new properties *do* seem to emerge. Novelty appears as a generic trait of nature and eventually may be generalized. What was once a fluke becomes common; what is at first an anomaly may become normalcy. Thus nature continually displays surprising new characteristics. There are always new things that appear under the Sun. Contingency, individuation, and historicity are generic factors encountered in the universe. But there are also new modes of coping and adaptation that appear, including homeostasis, natural innovation, and creativity.

According to the advocates of the theory of emergent evolution (such as Lloyd Morgan), such holistic qualities or properties appear at complex levels of organizations, and display unexpected rearrangement of existing entities. This can be seen at critical turning points in the emergence of new species, with the appearances of new appendages or capacities that did not exist before. These are unique to the members of a given species, which may eventually be generalized throughout a species by transmission to offspring. Advocates of this approach postulate that at some point the emergence of life was itself a "singularity," which appeared from nonlife; similarly for the appearance of conscious behavior. I am unwilling to simply assume the speculative thesis that the historical evolution of life is in a clearly demarcated straight line. *Homo sapiens* as a species survived, but others, such as *Homo habilis*, did not. For all we know, organic matter may have existed side by side with inorganic matter in various planetary systems, and organic matter *may* have always been complementary to inorganic matter. Many species with extraordinarily talented capacities have already become extinct. Thus evolution is not lineally progressive, and any emergents that we discover are not mysterious entities but can be given empirical explanations at the level in which they are observed.

This kind of inquiry is interdisciplinary; it tries to find analogous similarities across disciplines. A chief proponent was Ludwig von Bertalanffy.[2] The basic idea was that all phenomena may be seen as a web of interrelations of the elements or parts of a system. Such interconnected relations are found in electrical, physical, organic, biological, and social levels. These evince similar properties. General systems theory has not provided laws; it does point out similarities and analogies between fields. It points to generalizations that apply to all systems or fields, though the specific connectedness and relationships differ.

According to general systems theory, there are various kinds of systems, fields, or levels in nature in which special emergent properties or qualities appear. The following is a general classification of different systems, fields, or levels:

ORGANIC SYSTEMS FIELDS

atoms	family
molecules	tribe
cells	community
organs	societal institutions
organisms	culture
species	communication network
ecosystems	civilization

PHYSICAL SYSTEMS

subatomic particles
atoms
molecules
elements/chemicals
objects/bodies
moons
planets
solar systems
galaxies
universe, multiverse

We can classify emergent properties in terms of their complexity and scale: subatomic particles and waves (electrons, photons, neutrons, quarks, etc.), atoms (mini solar systems), molecules that contain more than one element, and cells (here nutrition, division, and replication emerge). Life is organic; its parts function in harmony in organisms. Cells form organs (heart, lungs, brain, kidneys) that are interrelated in internal systems (circulation, respiration, etc.) within the body; similarly for the parts of plants. Species characterize large numbers of individuals in social groupings (beehives, ant mounds, packs, schools, prides, etc.), extended human families and tribes, villages and cities, institutions and nation-states, social and cultural manifestations. We know that natural selection accounts for the evolution of species, but whether subatomic particles preceded molecules, cells followed, and more complex systems emerged—including solar systems, galaxies, multiverses, and the like—and how these stages emerged are open questions requiring theoretical explanations with some basis in observation and confirmation. To speak otherwise is a form of pure conjecture. These are insightful possibilities (or even probabilities), but we cannot invoke them on an untested timescale, even though simplest forms of life may be developed in the laboratory or in a primal soup on our planet.

Some theorists believe that creative processes are a pervasive dimension of reality. These advocates of "creative evolution" include Henri Bergson, who postulated an élan vital, or a special "life force." This thesis is highly questionable. Similarly for natural teleology: there is no empirical evidence for purposes or entelechies at work in nature, or for any future goals of the entire evolutionary universe, which is to be reached in a kind of progressive evolutionary advancement, as suggested by Pierre Teilhard de Chardin.

On the other hand, I think that functional explanations in biology are useful, but this is not the same as classical teleology. One needs to know the functions of different groups of cells (such as heart muscles), or of organs as part of the circulatory system, or of the lungs within the respiratory system, and the physical functions of all the organs in the body. This does not in any sense invoke design: an organism is not designed *a priori* (as fulfilling an architectural master plan); functions appear *a posteriori* and often by chance (according to natural selection). It is most unfortunate that the argument from design was introduced by Aristotle as part of his theoretical physics. It was a principle that explained how fifty-five unmoved movers were kept spinning. It was not to be used as a principle of creation. Aquinas and other theologians attempted to reconcile Christianity with Greek philosophy, so they read the Christian theistic God and his act of creation *into* nature. "Design" needed a divine designer. This is a specious argument by analogy with human designers on Earth, as David Hume pointed out. Once teleology disappeared from science, there was no need for a first cause or unmoved mover, or other confusing postulates. These do not explain the origin of the universe at all but only push our ignorance back one step, and, far worse, they only compound our self-deception into thinking that there is a hidden explanation, whereas there may be none at all.

Nevertheless, functional explanations have been useful in biology and medicine. The father of modern physiology, W. B. Cannon, introduced the concept of *homeostasis* to explain how the body functions. The term *homeostasis* literally means "to stand equally," and it describes how organisms need to maintain a degree of equilibrium or balance if they are to survive. He observed how many illnesses and diseases involved the disequilibrium or imbalance of body functions and how white blood cells are mobilized to combat infections. Thus diabetes occurs when the pancreas is unable to produce enough insulin to rid the body of excess sugars. Hypoglycemia may occur when there is insufficient glucose. Homeostasis can be graphically observed when the body is dehydrated. Similarly for the need to expel toxins into the bloodstream or increase white blood cells to combat infection.

There are mechanisms in the body that regulate normal levels of blood sugar, water, and various minerals and nutrients. Thus the organism attempts to regulate the internal environment of the body so as to sustain stable and constant conditions. There are multiple dynamic adjustments and regulatory mechanisms that function automatically below the level of consciousness or awareness. Cannon said that there was a kind of "wisdom of the body" (homeostasis) and that most of this was unconscious—though at times it may seep into awareness, as when one becomes thirsty, hot, and sweaty at the onset of dehydration and the organism seeks to quench its thirst and cool down. There are three control mechanisms: (a) the sensing of stimuli by receptors that send that information to the central nervous system; (b) a control center that sets the normal range, which, though variable, is maintained; and (c) effectors, which provide feedback. This suggests a kind of internal, unconscious biochemical intelligence at work. I use the term *intelligence* advisably—perhaps only metaphorically—yet one sees something like this throughout the biosphere. Thus natural selection is a form of "intelligent" selection in the sense that adaptive mechanisms and capacities evolve over long periods of time. There is no implication of design before the fact but only adaptive functions that are favorable properties that enable those creatures that have it to survive and reproduce and thus to pass it on to future generations of the species.

Some forms of life are able to adapt to changes in the environment. Thus cold-blooded reptiles do not maintain constant body temperatures, but are able to maintain life at a variable temperature. Equilibrium processes seem also to function in the external world. In an extended sense we see that an ecosystem maintains a kind of ecological balance, at least among the species that live in it. If a new species is introduced in an area, it may disrupt the life processes of existing species that come to count on other species providing a food supply for those who depend on it, such that its absence may spell disruptive disequilibria for the entire interdependent ecosystem.

I am always intrigued how the huge trees surrounding my house survive the severe winters of western New York. As autumn approaches, the leaves drop and the trees become dormant, withstanding heavy windstorms, though branches often crack or break off. In the spring, as the Sun appears, the trees re-bud and the leaves open up to convert chlorophyll into energy received from the Sun, to reduce CO_2 into carbon, releasing oxygen into the atmosphere. A new tree, planted in the fall, is able to adapt and adjust automatically by physical-chemical-biological processes at work.

Later, with more complex, higher-level organisms, forms of consciousness appear. A squirrel is able to collect and eat nuts and to flee menacing predators; a cat or a dog can respond to the commands of its human master and can learn to understand signals and behave accordingly. The growth of consciousness entails perception, conation, emotion, the emergence of cognition, and even intelligent behavior. Thus the instinctive wisdom of the body is heightened by the growth of a highly adaptive brain and the emergence of conscious rationality, which is conditioned in a cultural environment and in which language, signs, symbols, and abstract thoughts appear. The conscious brain (based on complex neural networks) is able to develop and extend still further the unconscious homeostatic wisdom of the body. A new dimension emerges when the human species is able to invent and apply tools and instruments that enable it to control the environment and to change it to suit human desires and purposes. Humans now can consciously provide shelter, food, and the necessities of life, so that we are able to survive and to thrive.

Functional explanations are essential in understanding society and culture. Political, economic, and social systems have functions to perform, and they can be best understood by seeing how they fit together and operate. Only explanations on emergent levels will suffice. I do not see how physics can tell us how to write a constitution or amend it; nor can chemistry and biology fully explain the economic system or tell us how to improve distribution, manufacturing, and productivity. Coduction seems to be relevant again here. To abandon the behavioral and social sciences entirely in the quest for a reductionist "holy grail" smacks of illusion.

Do we live in a contingent universe? We surely encounter contingencies almost daily. Are they endemic to existence, or are they sufficiently rare? Are they irreducible in themselves, or do we say that something is contingent simply because we do not know its causes, or the initial conditions under which an event occurs, with the result that we are unable to predict what will happen and when?

Contingencies may be accidents, chance events, misadventures, exceptions to the rule, freaks, unexpected or improbable, unpredictable, unique, or bizarre. Although all events are caused—in the sense that they do not pop into existence out of nowhere—we may not know which initial conditions or causes will operate at any one time.

In the real world very few events are absolutely certain or inevitable. More likely, many events are only probable or even highly probable, given normal initial conditions. But there may always be exceptions, and this depends on the scale or magnitude of a system, organization, organism, or machine.

David Hume assures us that we build up expectations based upon our past experiences—that the Sun will rise tomorrow, for example. Of course, it may rain or be cloudy, and we may not be able to see the Sun. But given what we know about the physics of our solar system, we can rest assured that Earth revolves on its axis every twenty-four hours and that it will continue to encircle the Sun through a predicted orbit. In fact, we can tell the precise time. All the planets rotate on their axes at divergent speeds. The Earth spins on its axis every 23.934 hours each day. This spin was faster in the past. Some six-hundred-million-plus years ago it took 21.9 hours, and it is likely to slow down in the future.

A relatively large system—such as our solar system—is fairly stable, and there are normal causal sequences that we can count on. There are other events on the surface of Earth that seem contingent—in the sense that new causal sequences may interrupt the normal flow of events. There are innumerable illustrations of this—I have already alluded to the earthquake and subsequent tsunami that caused a huge tidal wave in the Indian Ocean in 2004, but there are many others, such as hurricanes or tornados—severe weather disruptions in which countless events are happening simultaneously. The path of the storm may not be precisely predictable, and may have elements of contingency, in the sense that the exact course of the storm is not exactly predetermined.

The same phenomena are detected in the life world. A raging forest fire may sweep a wide swath of trees, and the intensity of their conflagration is dependent on the winds—their direction and whether they will die down. The brush and trees trapped at the center of the firestorm may not survive, nor will the animals, which are forced to flee for their lives and may not make it—all of this is quite independent of any human intervention exerted to redirect or squelch the rapid advance of the fire.

Similarly for events in the physical world, such as the meteors and asteroids that have impacted the surface of the planets—the geological terrain of the Earth and the Moon clearly show pockmarks. Does chance intervene at all, or is it simply the collision of two or more celestial bodies interacting and hence contingent?

If we examine the sky through a telescope, we are able to detect huge supernovae explosions in outer space; eighty-two were recorded one year, but this is a small fraction of those that occur, an infinitesimal number, given the trillions of stars that exist in the universe. These supernovae indicate the sudden implosion of a star, which becomes very bright (perhaps as bright as an entire galaxy) and in short order disappears from the lens. We are able to detect dwarf stars that have already burned out. Spectroscopic analysis of light emitting from the Andromeda Galaxy notes that it is shifting toward the blue, which indicates that it is approaching our own Milky Way and most likely will collide with our galaxy billions of years from now. All of the evidence in cosmology points to the continued expansion of the universe and an acceleration of the speeds.

So the classical belief that the heavenly bodies are everlasting and eternal, never coming into being or passing away, as we have seen, is mistaken. We live in a dynamic, pluralistic universe in which separate streams of causality are conflicting.

CONTINGENCY DEFINED

It is time to define contingency directly.

An event (E1) is said to be *caused* (C1) if certain *standard conditions* (SC1, 2, 3, etc.) are present.

1. An event is *contingent* if other causes or conditions or events intervene (e.g., a bolt of lightning strikes a house, a virulent virus is contracted from an insect bite, a flood inundates an area, etc.).
2. An event is *necessary* if and only if it must occur, without exception.
3. An event is *uncertain* if there are several conflicting causal sequences or events at work, and if it cannot be determined beforehand which will prevail.

* * *

Commentary on the above:

1. This seems to occur throughout nature, and on the basis of this we formulate causal regularities.
2. Contingencies occur throughout nature, contesting the uniform regularities formerly observed.
3. It is rare that we find necessary events in nature (as distinct from highly probable ones), because we are uncertain about the initial conditions present and the intervention of conflicting causes.
4. There are degrees of certainty and uncertainty and various levels of probability in nature.

Whatever will be depends on the interactions at any one moment of separate lines of causality, interacting, impinging, or crashing, and one line may supersede all others. A tree falls in the forest because it is laden with snow and heavy winds from a blizzard felled it; or a bolt of lightning splits it in half. A beaver is trapped by a fox that clutches its throat, prepared to devour it—two independent lines of causality battle for preeminence. The Mount St. Helens volcano (in the state of Washington) erupts, splattering the countryside and blotting out many life forms unable to flee in time.

One further consideration in dealing with contingencies is that many sciences draw on statistical data. Often that is the best we can do. We describe certain populations in probabilistic terms (as in quantum mechanics) that tell us on average what is happening by computing the average, the mean, or the median. We can thus depict in a general way the contours of the data under study. We may provide a graphic account of trends. The description of this graphic account tells us what values a population will evince. Sociological and psychological studies find a significant correlation at the .05 level, which is surely not a strictly causal relationship. This is important in those studies that discover many factors at work.

One objection offered to randomness is the so-called butterfly effect that ensues, even though we may see the role of a random series of events in nature. Thus, though a butterfly flits from flower to flower in what appears to be a haphazard way, still the sum of all random events apparently does have effects, and the fluttering wings are similar to eddies in pools of water, and thus they are part of a vast network of causal sequences. Accordingly, even in situations of apparent chaos there may be, or so we are told, causal consequences. Granted, but this suggests a kind of wavelike phenomenon, not easily subsumed under causal laws. Moreover, though the butterfly factor plays a role in the order of nature, it does not point to the interconnectedness of everything in a grid-like manner.

I submit that there are fairly isolated systems wherein internal causality operates. In the evolving biosphere, isolated breeding pockets of the same species may in time alter their genetic codes and thus become a separate species. The same considerations apply to isolated ecosystems separated geographically from others. What goes on in distant solar systems or galaxies may have very little or no effect on what happens in our solar system or on the planet Earth.

Thus, although physicists have developed four forces—gravitational, electromagnetic, and the strong and weak forces—this tells us very little if anything about other causal factors that apply to other complex systems and their interrelationships, especially in the organic and human spheres; and here new hypotheses and theories need to be introduced.

A good illustration of the need to develop higher-level laws to supplement physics and chemistry is within the field of neuroscience. We should try to uncover the neurological constituents of consciousness, perception, conation, emotion, cognition, and other psychological aspects of human experience (and/or of the experiences of other organisms). This means that the exploration of brain circuitry, the mapping of how the brain and neurological system works, and finding physical-chemical correlates and stimuli of psychological functions is vital. In a real and exciting sense, this form of research is on the frontier of knowledge, and it needs to be pursued vigorously. However, there are so many things that we still do not understand about the brain and how it functions, that this knowledge needs to be supplemented by concepts and hypotheses relevant to higher-order emergents.

The important proviso that I would include here is that there are *other kinds of explanations* that are also important. I will provide four illustrations of the need for higher-level explanations. The first is the groundbreaking research in psychology into the role of authority in obeying commands initiated by Stanley Milgram at Yale University in 1960–63 and carried out in a psychological laboratory. It underlines the easy willingness of so many people to follow commands even if it means inflicting pain on innocent persons. It illustrated the influence of authority in human psychology, and especially its role in closed societies and dictatorships. Another illustration from psychology is the use of rational-emotive therapy by Albert Ellis, wherein the analysis of both cognition and emotion by the patient in cooperation with the psychotherapist seems to be a useful form of therapy without the need for pharmaceutical drugs or the doubtful use of psychoanalytic techniques.

A second field of inquiry in which reductionism is insufficient is economics. Is it a science? Economists serve many functions. No doubt they have a practical advisory role in business and government; they seek to develop policies that will maximize profits in commercial firms. They may recommend to a government how to stimulate the economy or develop equitable tax policies. Accordingly, economics is an applied policy science. A more basic function of economics is how to understand economic behavior. Adam Smith's classic *Wealth of Nations* attempted to ascertain how to achieve a prosperous society in market economics, as did the inquiries of David Ricardo, John Stuart Mill, John Maynard Keynes, and other economists.

One classic criterion of economic change is the relationship between supply and demand. Price is determined as a function of choices exercised in the marketplace by rational producers and consumers, two abstractions that may not actually be fully exemplified in society, for advertisers titillate taste and predilection and emotion, as well as reason, which undoubtedly play important roles in supply and demand. This can be seen when speculation runs riot and "the madness of crowd" psychology intervenes, as in the historic tulip mania that erupted in the Netherlands in 1637, where the price of tulip bulbs soared, encouraged by a speculative frenzy. Be that as it may, I fail to see how monitoring microevents in a great number of consumers and producers by itself can possibly be a substitute for dealing with economic behavior in the actual world of real women and men. Adam Smith's analysis is useful, and it needs to be supplemented by principles introduced by other economic theorists, but these are on the level of social interaction.

A dramatic illustration is the rise and fall of stock markets all over the world. Indeed, we may ask, what is a stock market? Can it be reduced to the sum of microevents in the brains of human beings without losing what it is? Obviously an account of consumer satisfaction is important in its own terms.

Hence I submit that this area of behavior is simply not reducible to microevents without losing the subject matter entirely. Is there not a kind of qualitative phenomena that has emerged, which must be dealt with on its own level?

One can ask, from another vantage point, what is the "corporation"? Is it a legal entity? Does it not have a life of its own, with a history, traditions, structures, and tests of productivity and success? Do we not have to deal with the Chevron Oil Company on its own terms, without explaining it away? Thus the agenda of reductionism is in danger of losing its distinctive subject matter. Of course a corporation is made up of human beings, past and present, yet its interactive behavior and the principles of its organization seem to have their own existence over and beyond any one person.

A third example of the limits of reductionism is from evolutionary biology. One may ask, how does Darwin's theory of natural selection explain the origin and/or extinction of species where there are many macrofactors at work: differential reproduction, adaptation, the struggle for survival of a species, and so on? In other words, the biogenetic principles of evolutionary theory—as change throughout time—exist at a more complex biological level, and the principles that were introduced by Darwin cannot be reduced to the "selfish gene" alone, but the ability of individuals to survive, the struggle with other individuals competing for females and the opportunity to reproduce is essential in the process, not simply the description of microevents. Mutations, of course, have a vital role to play in explanation, but not by ignoring other factors at work.

A fourth rather large area to consider is the realm of cultural patterns and sociological institutions. Human behavior clearly may be described in cultural terms, because human beings are involved in distinctive cultural activities. In one sense, according to anthropologists such as Alfred Kroeber and Clyde Kluckhohn, culture in general refers to the accumulated historic treasury of human creations: books, works of art, buildings and monuments, linguistic systems, religious tradition, morality, science built up through the ages, and so on. A culture thus may be viewed on a larger scale, encompassing many individuals and their ideas and values *and* the interactions of these as played out over periods of time and geography. The Renaissance thus is a sweeping term in cultural history, appearing in cities across Europe. To attempt to reduce it to its physicalist elements would be unnecessary and ludicrous.

One may ask, how should we deal with sociocultural institutions such as nation-states (France or Spain), religious systems (Islam or the Roman Catholic Church), symphonic orchestras or musical organizations (the Cleveland Orchestra or the Metropolitan Opera), museums of art (the Louvre or the Tate Museum), or great universities (Princeton, La Sorbonne, Cambridge, or Stanford)?

Now, such institutions are no doubt made up of individuals in the present and the past. They each have traditions, structures, and personnel; more important, they have goals to fulfill and criteria to evaluate their performances. These simply are not reducible to physics, chemistry, or biology—though these sciences may have important knowledge to convey about any number of skills that individuals in the institutions possess and/or need to fulfill. Thus, I submit that multicausality, that is, pluralistic conceptions of causality, is relevant.

TRANSACTIONAL QUALITIES AND RELATIONSHIPS

It is necessary at this point to discuss the ontological status of emergents as *transactional* qualities and relationships. From the standpoint of the physicalist, all that exists are material bodies, defined by their properties or qualities. Primary qualities refer to the mass, weight, size, and shape of a body. Secondary qualities refer to the powers that things have to cause experiences in sentient beings—color, taste, smell, sound, and touch. Tertiary qualities refer to the complex relationship that exists among life forms. They are dispositional qualities that function within fields of interaction. Both secondary and tertiary qualities are emergents—a function of stimuli (e.g., the color of a flower or its fragrance) and their appearance in a perceptual field.

Transactional qualities involve the more complex relationships between objects and their properties and sentient beings, and especially between sentient beings themselves. A graphic illustration of this is sexual arousal between two organisms. For example, the beautiful display of the male peacock or, in some species of bird, the dance of the male to attract the female. Between humans this is known as romantic love, the source of affection, poetry, and melody; the fragrance and beauty of touch and orgasm. What is the ontological status of sexual behavior? Obviously it is a *tertiary* quality, even between same-sex couples. It is a coupling relationship leading to foreplay and orgasm, a quality that emerges in a relationship between two or more members in a sexual liaison.

Conversely, why do males—gorillas or caribou or bulls—battle ferociously with other males for possession of the female(s) and the opportunity to reproduce? Their antipathies are rooted in the raging testosterone and in many species eventuate in fierce combat. Competition is a transactional quality in the battle for dominance, an emergent display that appears, involving two or more males who battle for dominance and conquest.

Another illustration of a transactional emergent is parenting, the instinctive behavior of a mother in many species to eat the placenta and lick the baby offspring clean. This allows the infant to suckle, protects it from danger, teaches it to fly (if a sparrow) or waddle and swim (if a duck). These emergents are relational qualities. If one views a mother with eight ducklings following her, the quality is in the behavior of the group.

A huge flock of birds in flight, coordinating their direction, is splendid to behold. It is not a collocation of nodes of atomic particles in air but is rather a field of interaction of hundreds or thousands of birds in flight. Similarly, a herd of elephants with their baby elephants protected from predators will trek hundreds of miles together in times of drought to seek water. And the offspring look to the parents for nourishment and help, so a form of filial devotion emerges during the period of upbringing and tutelage.

A beehive, hornets' nest, or ant colony is a rudimentary social system, with a queen bee or a hornet or soldier ants all working together to protect the hearth and home and reproduce countless numbers of offspring to survive. If this analysis is extended to the human domain, it is open-ended because over and beyond the biological dispositions and instincts of the humans emerge qualities that are psycho-socio-cultural. These are truly open-ended and amenable to incredible diversity and richness.

It should be apparent that there are *relationships* between two or more individuals that can be understood only on their own terms, without a reduction to the singular individual alone. Perhaps the best illustration is sexuality itself, for one could not possibly understand the male's sexual organs (the penis to insert and the testicles to produce sperm) and the capacity to ejaculate unless one understood their relationship with the female's sexual organs (the vagina, receptive to the penis, and the ovaries, to produce eggs). It is in the interaction between two sets of plumbing that one can figure out their functions. Of course either party may masturbate by self-stimulation, and the penis may be inserted into the mouth or rectum or goat's vagina (as farm boys are allegedly wont to do). Yet it is the transactional functions of both sets of sexual organs that enable us to comprehend their natural biological functions.

The same thing may be said of the family unit; the female's breasts are to provide nourishment to the infant. And the male often assumes the role of protecting the brood. This applies to all species—we observe ducks feeding the ducklings and then teaching them to peck for food and to swim in a pond or stream. Parenthood is a relational quality, where the female generally plays a dominant role; similarly for the hornets' nests (with structures like honeycombs), in which a large number of worker hornets are pressed into service, fulfilling a variety of tasks. Incidentally, hornets and bees are interrelated to the flowers and plants that they extract nourishment from and in the process carry the spores of insemination far and wide. All of these are transactional qualities that have emerged. They are interdependent realities.

If we extrapolate further, the same transactional relationships apply to other members of the family (brothers and sisters and grandparents [who may be dependent on younger members for feeding and protecting them from dangers]). We may take the process one step beyond to the tribe, where a division of labor evolves, or to the village and other more complex fields of transaction, for example, a highway system, or a reservoir and a network of pipes to distribute water to farmers in the countryside and to residents in urban locations.

Thus transactional relations are new emergent fields, which manifest a reality on their own level of interaction, and these interdependent sociocultural entities are as "real" as the isolated or singular human by itself.

The family unit was rudimentary for the survival of primitive humans. Similarly for the extended family and tribe bound together by the need for human protection, given the long period of time needed for nourishment and maturation of infants and children. Humans huddle together to ward off threatening wild animals and other tribes, deprivation, hunger, or climate changes.

Food gatherers and hunters work in unison and later become members of a consanguineous clan, tribe, or village. Systems of social relationship thus develop to fulfill common human needs. These are tertiary relational qualities that have emerged. In time, cultural practices appear: languages for communication; tales passed down by oral traditions; crude tools, later perfected, used to hunt, fish, carve, or scrape. Eventually the discovery of fire; the invention of the wheel, cart, or shelter so that the use of technological tools is learned and transmitted to future generations, enforced by tradition; the emergence of religious rituals, shamans, and medicine men is part of a division of labor.

What is the ontological status of the original social groups, with a tribal chief, hunters, those who gather wood for a fire and build shelters, those who nurse and care for the children, and those who defend the tribe from marauders? There is implicit in this a rudimentary division of tasks and a social structure. These institutions are engrained in social habits and the concomitant systems of beliefs that support them, and they are *cultural emergents*. Physicalist reductionism tells almost nothing of these evolving systems, which accompany biological evolution; yet sociocultural changes can radically alter human behavior—as new tools are discovered, they transform who and what we are as agricultural technology and domesticated animals for work, milk, and meat enable humans to establish permanent settlements. The discovery of how to make bread, wine, cheese, and olive oil solves the problem of preservation of food and enables it to be available all year long. Urban dwellers develop new divisions of labor—as Plato pointed out in *The Republic*. This led to more complicated social systems—an army; a set of laws; a political class; teachers, poets, and writers; beauticians and masseuses; chefs and tapestry weavers; and all of the fineries of civilization. These are tertiary qualitative emergents, which are *real*, indeed *more real* and important than the objects and practices of their primitive forebears. They are properties of social systems and cultural patterns, transmitted from generation to generation. A class system emerges, along with property and wealth, trade and commerce, landowners and serfs, philosophers and statesmen.

None of these phenomena can be understood alone by knowing the neurological networks in the brains of individuals or even their psychological dispositions and capacities alone—though this knowledge is essential in a full account of who and what we are.

Of crucial significance in understanding society and culture, institutions and nation-states, churches and stock markets, universities and art museums, governments and political parties is that we cope with different subject matters on their own terms. We need to understand ancient Babylon and Mesopotamia, Chinese culture and Japanese samurai warriors, African art and European culture, North and South America, Paris and New York, Buenos Aires and Sydney, Australia. Here we are dealing with a new level of inquiry, human affairs, and human history and the role of individuation in historical changes. And this has brought us beyond physics and chemistry to the behavioral and social sciences and to the complex world of culture and civilization.

ACT SIX
INDIVIDUAL ENTITIES AND THEIR HISTORIES

INDIVIDUATION

From our discussions of the various sciences, it is evident that individuation is found throughout nature. It is a *generic trait* that applies to countless concrete entities, objects, things, processes, and events. It applies to the biosphere, the physical universe, and human affairs. There are so many different particular individuals discernible in nature. These include planets and moons, galaxies and asteroids, continents and islands, rivers and streams, parrots and geese, palm trees and rhododendrons, horses and bulls, human beings and chimpanzees, social institutions and cultural artifacts, atoms and molecules, electrons and photons, cells and organs. One cannot easily explain these away. They are *givens*—an incalculable number. Moreover, they are discernible by their diversity. These are the rudimentary bedrock constituents of the natural world: no two snowflakes, grains of sand or diamonds, girls or boys, doves or kittens are exactly alike. Much the same as no two human beings have the exact same fingerprints! Thus a vast range of such individual concrete objects or entities interact with other objects and entities encountered in the world.

Aristotle used the term *ousia* (substance) to indicate things that exist, though they could be classified into different genus classes, each a unique thing. I would describe these diverse things as products of contingent influences; *chance* intervenes here, for although they may be developed through common processes of creation, evolution, growth, and development, nonetheless there is a kind of brute facticity that exists. This includes everything from pebbles on the beach, rocks and boulders, to ants and earthworms, sheep and lions, flowers and plants, pens and computers, persons and institutions, planets and stars.

To this we need to add the *properties* of individual objects or things and the *events* that they undergo. It is difficult to imagine the existence of any properties or qualities without their base; similarly for *relationships* behind these individual objects and their qualities. They possess a kind of objective structure, yet in spite of this there are differences so that they may not be exactly alike. So individuation is a basic category of what it means to be. There are also discernible systems of interconnected and interacting fields, forces, and events. A solar system has planets circling a star, and these within a galaxy or cluster of galaxies; similarly for schools of fish and flocks of birds, rainforests and coral reefs, beehives and ant colonies.

Another generic trait that manifests itself is the appearance of *idiosyncratic individuality*, particularly in the life world of biosystems. This applies notably to human beings but is also found in other species. I have noted that in a litter of puppies, each seems to differ; one may be frisky and spirited, another shy and laid back; one may be combative, another meek and passive. Idiosyncratic qualities are everywhere present. Even in fraternal twins with similar genetic stock there is a wide deviance in tastes and attitudes, beliefs and values, needs and interests. Although individuals may have the same or similar environmental backgrounds, they may exhibit different personality traits. My former student, Sam, has a voracious appetite and will eat almost anything served to him. His brother, Hank, is a finicky eater, turning up his nose at everything that does not fit into past patterns; he does not like vegetables, but he will eat a gallon of ice cream. He resists any food that is new—such as crab cakes or lobsters for the first time; whereas Sam is receptive to any nouvelle cuisine and will devour it with gusto. He is a hardworking student and a great baseball player. His brother plays soccer with consummate skill and studies science.

We learn that individuality is a pervasive trait of human life: some individuals like classical music—Bach and Beethoven or Mahler and Bartók—others may loathe it. Some adore jazz, others rock and roll; some will read voraciously, others never crack a book. Margaret loves to go dancing all night long; Sharon prefers quiet evenings at home. Seymour loves outdoor activities; Bob prefers to spend time in the library. Jim is a conservative Republican; Sandra is a left-wing

Democrat. Some personalities become Libertarian taxi drivers, others beauticians. Some join the army; others prefer to stay in graduate school and pursue research.

Thus, though we share many values and ideas with others, we also differ in our attitudes and convictions. How many political parties are there in France? The answer is sixty million, one for every French citizen! Charles de Gaulle is reputed to have said that France is ungovernable because it has a different cheese and wine for every province or region!

Some cultures encourage individuality and pluck and tolerate idiosyncratic diversity in tastes, beliefs, and lifestyles. Others try to dampen personality differences and instill conformity and obedience in individuals. Some people are turned on by highly charged erotic sexual activities; others will only have an orgasm using the missionary position. I submit that individuality is a product of contingent events that have influenced who and what an individual will become. This is due to their genetic stock, the fetal influences in the mother's womb, and the imprinting and conditioning of the environment, especially in the early formative years—and how they react to these stimuli. Although human beings share so many needs, capacities, and pleasures and engage in common activities, these are relative to the contingent influences on each person, which stamp him or her as unique. This involves the ability to learn from experience and to change behavior—within limits—in the light of reason and whatever level of understanding a person (especially a child) has attained.

I was bowled over to read that this also applies to even the simplest forms of life, including *E. coli*, a species of bacteria that exist harmlessly in every person's intestines—we each have trillions of them;colonies of genetically identical *E. coli* (*Escherichia coli*).[1] According to science writer Carl Zimmer, scientists who have studied *E. coli* thoroughly find it a striking fact that the same principles of life found in *E. coli* are found everywhere in living things. The French biologist Jacques Monod has said, "What is true for the *E. coli* is true for the elephant," and we may add the sea cucumber, dolphin, and human being as well. In spite of the great diversity of life forms that we encounter, life is astonishingly uniform; for example, *E. coli* use both RNA and the same genetic code to transfer it into protein (flagellum) and all encode genes with DNA. Thus the world of the microcosm reflects the biosphere in general, and the science of microbiology drawing upon physical chemistry has unraveled the basic building blocks and processes of life. This supports the reductionists' agenda, yet there is much more to the story.

The study of *E. coli* reveals other remarkable features that are similar to what we find in other forms of organic life. *E. coli* engage in sex, cooperate with other members of their species, communicate with them; they wage warfare, make chemical weapons, and also may even sacrifice themselves on occasion. As for the colony, Zimmer maintains that *E. coli* have social lives. But they also display a stunning form of individuality. In this sense *E. coli* behave like a mob of individuals who behave differently, even under identical conditions. If two *E. coli* are swimming side by side, one keeps spinning its corkscrew tail while the other gives up. If a colony is fed lactose (the sugar in milk), some will turn away while others will slurp it up. *E. coli*, says Zimmer, are unpredictable; one clone may produce many copies of protein, while another will produce zilch. Living things are not run by their genetic programs alone; thus even microbes with the same genetic network can lead to different outcomes. *E. coli* are able to incorporate genes from elsewhere into their bodies; *E. coli* thus mutate. They seem to engage in bioengineering.

The fact that basic biochemical regularities are found in *E. coli* and all forms of life nonetheless leaves room for environmental factors, historical challenges, and the expression of individual differences.

Identical twins brought up in the same family can have radically different tastes and beliefs, because environmental influences are not exactly alike. It is reported that even animals cloned from the same cell develop different personalities, because the total sets of experiences in life are different. What is often overlooked is that individuation also applies to solely physical objects and events. This applies, as we have seen, to planets in our solar system and to the solar system itself, and it applies even more so to the cultures and societies of today as well as in the past.

HISTORICITY

An essential source of understanding anything in nature is knowledge of history. The differences that all objects manifest—whether organic or inorganic—may be analyzed by reconstructing their pasts. This applies to birds and bees, beetles and trees, as it does to planets and galaxies. It also applies most directly to understanding personal biographies and the changes observed in social and cultural institutions.

This means that we need to view nature and life in terms of change, process, and flux. Newtonian science looked upon the universe as fixed; its elegant mechanical laws enabled us to predict the orbits of planets around the Sun as well as the behavior of bodies on Earth. In the nineteenth century there was a refocusing on historical change; Hegel and Marx took historical events as the center of intellectual attention and attempted to explain the march of reason through history. Thus changes through time needed to be unraveled. The focus on historical change made possible the emergence of Darwinian evolution, for species changed through time; and what they are today is different than what they had been in years past, as the fossil record clearly demonstrates.

Change over time involves two kinds of explanations: first, the *origin* or beginning of an object or event, person or country, species or field. The birth of a child or the creation of a natural object, such as a statue or a road, or a great social institution such as the US Constitution—second, we also need to understand the *process* by which each changed, was developed, grew, and was modified, or, conversely, how it fell apart and was destroyed. Aristotle called this the "efficient cause," how something came into being and/or was disassembled—as the eruption of a volcano, which eventuated in a mountain, or the collapse of a building by an earthquake, or the destruction of a country such as Carthage.

The accounts of historical changes in society are essential if we are to understand a nation. Let us look at the Americas, which were first settled by Mongolians, who came from Asia over the Bering Strait to North and South America. In time hundreds of tribes and civilizations developed. We can trace their histories by examining the remains of monuments and buildings such as the Inca civilization of Peru. Historians can describe their works of art, their methods of farming, how they buried their dead. Historians have described the hundreds of tribes from New England to Florida and California and their diverse languages and customs.

The first Europeans to come to the Americans were Leif Ericson and the Vikings in the eleventh century. This was followed centuries later by larger groups of European explorers and colonists from England, Spain, the Netherlands, France, and Portugal. By 1776 the English colonists revolted against the British Crown, declared independence, drafted a constitution, were led by outstanding statesmen (Washington, Jefferson, Madison, and Monroe), and so the United States of America was born. And we can trace its rapid growth from three million inhabitants in the eastern and southern colonies to the conquering of the frontier, subduing the Native Americans, until forty-eight states were established and eventually fifty. This was accomplished by opening the gates to immigrants from all over the world, by a brutal civil war that ended slavery (with its heroic figure Abraham Lincoln), the building of railroads, and rapid industrialization. By the beginning of the twentieth century, America had grown to one hundred million inhabitants (Theodore Roosevelt led the new America). The status of the United States of America as a world power was secure; it fought Spain for the Philippines and entered World War I to save England and France from Germany and the Habsburg and Ottoman Empires, and, according to former Princeton scholar and president Woodrow Wilson, "to make the world safe for democracy."

America's economic, scientific, and technological advances astonished the world with provocative skyscrapers and the country's powerful navy. And again, after the devastating Great Depression of the 1930s, the United States intervened in Europe against the fascist states of Germany and Italy, and in Asia it contested the rising power of Japan. The post–World War II era saw the rise of two superpowers: the United States and the Soviet Union; and a Cold War between democratic capitalism and Soviet-style communism resulted in the collapse of Russia and supremacy of free-market economies everywhere, including Red China, whose revolution frightened the West but whose economic power invited investment and trade. There were wars between North and South Korea and North and South Vietnam, and the gradual disappearance of global confrontation. The country also saw the growth of the United Nations, at first as an effort of the victors of World War II to establish a system of collective security (which was of limited success), and of the Universal Declaration of Human Rights. By the year 2000 the United States had grown to some three hundred million people; Europe attempted to heal its old, festering nationalistic wounds and to develop a common market and European parliament. Indeed, Europe's total economic power was roughly equivalent to that of the United States, but emerging to contest both was the economic power of Japan (by now the second-largest economy in the world), South Korea, and the growing power of China as an industrial and technological nation, with other emerging nations in competition, notably Brazil in Latin America and India in South Asia.

Standards of living are rising everywhere, with the green revolution in agriculture forestalling famines and the growth of medical technology (antibiotics, surgery, better sanitation) leading to improved conditions of living for ordinary people. There are of course exceptions: problems of hunger and disease still persist in Africa, Asia, parts of Latin America, and India and Bangladesh in South Asia; and prosperity is confronted by economic dislocation and recession, unemployment and continued population growth, strained resources and polluted streams and atmosphere, with global warming an ever-present threat.

On the sideline stands Islamic fundamentalism, which objects to modernization, the liberation of women, democratization, and technological growth—as if a whimper by fanatic terrorists can stop socioeconomic progress and the eventual demand of its young people to share in the prospects of freedom and a better standard of living—all inspired by the World Wide Web. The world is gradually becoming conscious of the need for new global institutions to deal with financial crises, the depletion of resources, global warming, and the need for transnational cooperation.

This brief global tour of the United States and the world is meant to show how historical analyses are vital—change is ongoing, and no sooner is one crisis met than two more emerge. There is no escape from the continuing onslaught of new problems. Retrospective historical analyses of political-economic-sociological, scientific, ideological, and cultural changes are therefore essential if we are to understand any one area of the world.

BIOGRAPHY

The same thing is true of our understanding of individuals. How are we to explain any person, who and what he or she is, without a detailed biographical account?

Consider how many biographies have been written about Abraham Lincoln: the story of the unlikely rise of a poor boy born in a log cabin on the frontier, self-educated, who rose to greatness as the president during the Civil War and who died a martyr (or traitor, depending on whether you're a defender of the North or the South). Lincoln had a wife, Mary Todd, a shrew who hounded him with vitriol even after he became president. Of the hundreds of biographies of the man, which was the true account? One biographer, C. A. Tripp, claims that Lincoln was gay or at least bisexual and that he shared a bed with young Billy Green, a frequent bunkmate of Lincoln's in the 1830s in Salem. A second relationship with a male friend went on for several years. This relationship could still be surmised from the letters of guilt and affection from his alleged paramour of the 1840s, Joshua Speed of Springfield, Illinois, who likewise shared a bed with Abe Lincoln for several years. Tripp observes that poverty did not require the longtime bed-sharing. There are also well-documented accounts of Lincoln's trysts in the White House with Captain David V. Derickson; both were seen in nightgowns together when Mary Lincoln was absent. The greatness of a man as a public figure is of course quite separate from his private escapades.[2]

Many consider Thomas Jefferson a semidivine figure as a Founding Father of the American Revolution and author of the Declaration of Independence. Yet his long-standing affair with Sally Hemings, his beautiful mulatto slave, was common knowledge in his day. He took her to Paris with him when he was ambassador.

James T. Callender, a political journalist who had once been an ally of Jefferson, wrote in 1802 that Jefferson had kept a concubine, Sally Hemings—his own slave—and that he sired several children with her. Much was made of the story by Jefferson's Federalist opponents in his day. Jefferson as a rule did not respond publicly to personal attacks on him, though the story was well known in the nineteenth century. Defenders of Jefferson said that his nephews Peter and Samuel Carr might have sired the children of many Monticello slaves, hence their resemblance to Jefferson.

A special research committee of the Thomas Jefferson Foundation reexamined the evidence, including a DNA study of the descendants of Sally Hemings. They concluded in 2000 that there is "a high probability that Jefferson fathered Eston Hemings" and that he most likely "was the father of all six of Sally Hemings' children."[3] No one is perfect, least of all great political leaders!

Many great historical figures were all too human. Benjamin Franklin, the great sage of Philadelphia, was a ladies' man who left a good part of his estate to his French mistress. Bill Clinton was nearly driven from office for lying about his affair with Monica Lewinsky. "I did not have sexual relations with that woman!" he insisted. What he meant was that they did not lie down, but there is substantial evidence that she performed fellatio on the president while he was standing up and that some of the sperm was found on Monica's dress.

My dowager French mother-in-law, then in her nineties, once asked me, "Tell me, why is the press so furious with Bill Clinton? Don't most heads of state have affairs?" she asked. Look at Louis XIV and his mistress, or President François Mitterrand of France, whose mistress lived with him in the *Élysée* (the French White House) when he was president. When Mitterrand died, his legal wife and two sons were photographed at the funeral gravesite, and next to them were his mistress and their daughter. This was published in the French press and surprised the French public, because it was the first that they had learned about his private life, something the French prefer to remain private (unlike *les Americaines*, who wish to tell all). Didn't President Kennedy have many women (allegedly including the actress Marilyn Monroe) spend the night with him at the White House, while, according to rumor, Jackie Kennedy had a torrid affair with the billionaire shipping magnate Aristotle Onassis? No, I said it was not the affair that Clinton's political enemies were agitated about but the fact that he perjured himself. "Qu'este que c'est 'perjured'?" my mother-in-law asked. The fact that he lied about it, to which she replied, "Who is going to tell the truth about an affair? Doesn't every public figure lie about it?" "Les Americaines, quelle naïveté!" she remarked. Shortly thereafter Nicolas Sarkozy, the newly elected divorced president of France was shown in the daily newspapers as having a torrid affair with the Italian model Carla Bruni. He promptly married her. So French public opinion is not as blasé as it used to be.

In any case, the point is obvious. If we are to understand any historical figure, biographical and autobiographical accounts are deep sources of information. To understand how and why decisions were made, we need to know a great deal about the personal lives of the men and women who made them, not simply their sexual peccadilloes but also their desires and passions, ambitions and jealousies, admirable and detestable characteristics.

I learned so much about the Soviet Union by reading three biographies by Dmitri Volkogonov (a former Soviet colonel general whose father was imprisoned by Stalin) about Stalin, Lenin, and Trotsky, three men who made the Russian Revolution.[4] Having read and studied so much about communism in the Soviet Union, I came to the conclusion that the 1917 revolution, which brought Lenin and the Bolsheviks to power, was betrayed by Stalin and the gulag that he created. How and why this happened could be unraveled in part by a diagnostic historical analysis of the lives of the three leaders. Lenin was the idealist who nonetheless first used terror to free Russia from the czarist regime and to solidify his power. Trotsky, founder of the Red Army, was forced to flee Russia because of his dispute with Stalin. Much later he was murdered in Mexico on orders from Stalin. Stalin was the man of steel, the ultimate party functionary who seized absolute power and liquidated any opposition to his authority, including his former idealistic Bolshevik comrades and millions of innocent Russians. The tragedy of Soviet communism and the shattering of ideas is a sad though eloquent commentary on the fate of utopian visionaries.

THE USE OF CIRCUMSTANTIAL EVIDENCE

Is the historian an applied scientist or a literary artist? Undoubtedly he or she is both. Historical biographies are often laced with great literary metaphors so as to give the flavor and feel of a person and era. A powerful work presents metaphorical poetry, as did the popular biographies of Alexander Hamilton and Samuel Adams. But clearly any historical account that is worthwhile needs to be impartial and objective. Its descriptions should be grounded in the facts, with precise dates, times, places, and sequences of events carefully recorded. These data should be based on the empirical record—birth and marriage certificates and deeds of sale testified to by the town clerk, dates listed on memorial slabs at the cemetery or inscribed on buildings and monuments, and more. Another source is the written account of reliable newspaper journalists and the authors of articles or books. Still another source is the testimony by eyewitnesses of such events. All of the above may be tarnished by sloppy records or biased journalists and historians with axes to grind. Especially questionable are uncorroborated eyewitness accounts. Much of this may be based on prejudicial testimony, second- or thirdhand accounts, which may be garbled in the retelling of an urban legend or rumor. Unfortunately there may be the inclination of some witnesses to read into an event their own interpretation. There may be willful, though perhaps unconscious, distortions rendered. If you ask six people at an auto accident, "What happened?" you may get six different views, depending on where the person was standing, or the personality or demeanor of the drivers. Of course, we need to rely on eyewitness testimony, but we must be assured that it is dispassionate and objective.

Also we must take into consideration the question of *when* the event happened. If it occurred yesterday or today, this is far different from events in the past and the problem of accurate memory. As time goes by, past events may be blurred and confused. Memories become faded, and we need to ascertain the actual facts of the case. Given all of these caveats, obviously historical accounts should be based on accurate factual evidence, either of the written documented record or of reliable witnesses.

In many cases, the evidence may be tarnished, yet we attempt to reconstruct as far as we can a truthful narrative. In most cases we need to rely on *circumstantial evidence*, and we need to fit it together. The key tool is the hypothesis. The investigator is like a detective or sleuth, trying to ascertain who did what, where, and when. The best illustration of this is Dr. Joe Nickell, who has a PhD in English but who has turned out to be one of the most important paranormal investigators in the latter part of the twentieth century. I am not denigrating Houdini (earlier in the twentieth century) or the Amazing Randi, a great magician and showman who has done topflight investigations, but Joe Nickell has devoted decades to careful, dispassionate investigations, assuming neither the posture of the wide-eyed "true believer" nor that of the "true unbeliever"—the dogmatic skeptical naysayer who rejects a claim out of hand prior to investigation.

For example, Nickell began an investigation of the so-called Lake Champlain monster with an impartial attitude. There have been numerous eyewitness accounts of a long, slithering creature in Lake Champlain, New York, over the decades. People of unquestionably good character claim to have seen it slinking along, showing one or more humps and a long neck above the water. There have even been photographs taken. Every few years popular journalists decide to play up the sightings, and TV channels devote special programs to this mystery.

Nickell's investigation began with the data of eyewitness accounts, too numerous to be dismissed or to be characterized as gullible nincompoopery; his inquest was for an explanation of the sightings. Something obviously was seen. What are some of the interpretations? A freshwater sea monster? If so, there would need to be more than one—a breeding population (unless it is a Methuselah-type monster). Have any babies been hatched or ejected? Have there been any underwater sightings; any collisions with bathers, canoes, or motorboats? None, as far as Nickell could tell. Has anyone been molested or bitten by this alleged monster? No such reports. The sightings have mostly been from afar. So the question is, *why* do people claim to see something? Is it like a mirage or misidentification? Several hypotheses have been suggested: First, loons have been frequently sighted in the vicinity. A loon is a long-necked bird that may have been in the lake and whose neck looks like that of a "sea monster." Second, otters often swim in line in the lake and so simulate a single, multi-humped monster. Third, floating logs could be mistaken for the "monster." Fourth, observers may be seeing a giant sturgeon. The answer may be all of the above or none; namely, perfectly prosaic explanations of natural phenomena, colored by an urban legend that encourages people to believe that there is a Lake Champlain monster and the insistence by observers that "I saw something that looked like one!"

* * *

I wish to reiterate that there is no sharp dividing line between physics and astronomy as theoretical sciences and the applied sciences of engineering, geology, meteorology, climatology, and so forth. It is one thing to have general physical-chemical laws; it is another to reconstruct what happened and why in concrete cases. As we have seen in Act Four, the natural sciences also deal with *historicity* and *individuation*. They attempt to understand the origins and historical changes in the planets and moons in our solar system. Thus, the history of the Earth, Mars, Venus, Jupiter, Saturn, the Moon, and their relationships to the Sun and also to the entire solar system within the Milky Way and its relationship to other galaxies in the universe is essential if we are to understand nature. The adequacy of causal regularities, such as the theory of gravitation, is constantly tested in different physical contexts: on the microlevel, in understanding physical objects in everyday life and calculating their movements, in the solar system, and in the galaxies beyond ours.

Similar applications apply in the biosphere, as we have seen in Acts Two and Three, in understanding how natural selection operates in explaining the evolutionary descent of species. Similar applications apply in biochemistry and in the understanding of the functions of biological systems.

All of these considerations apply as well to human affairs, where historical reconstructions of individuals, communities, nation-states, cultures, and civilizations are especially relevant. It is to this area of inquiry that we now turn.

ACT SEVEN
CONTINGENCY AND CONFLICT IN HUMAN AFFAIRS

COEVOLUTION: BIOGENETIC *AND* CULTURAL

The human species lived in small hunter-gatherer groups for most of its existence. Humans are hardwired to certain forms of behavior that developed over long periods of evolutionary history. As we have seen, *natural selection* emphasized certain forms of behavior that proved to be favorable for survival and that were transmitted to future generations.

The infant is not abandoned at an early age, for there is the nurturing care of parents, real or surrogate. They continue to provide a safe and predictable environment for their children. They feed and protect them against predators until such time as they are able to fend for themselves, walk, seek nourishment, discover shelter, find a place to sleep and rest, and learn how to flee or defend themselves from threats in the environment. This is true for mammals—birds, kittens, puppies, or calves—though most need a relatively short time of parental care. Birds will tend their chicks in the nest and bring food, and they await the first efforts to fly on their own and become self-sufficient. For elephants, dependency is for an extended period, and for humans, even longer. Another dynamic factor in the evolution and enhancement of human life has been the eventual emergence of cultural and social institutions.

Interestingly, fragmentary remnants of hunter-gatherer groups still exist, though they constitute only very few isolated peoples, some in East Africa, a few in Papua New Guinea, the Amazon, and the Arctic. Hunter-gatherers were the dominant mode of existence of the genus *Homo* going back two million years. Humanity survived by hunting for beasts of prey and/or as food gatherers—as do all other species living off the land and traveling to find food such as plants and berries, or to kill and consume prey. A recent *National Geographic* study of the Hadza people in Tanzania, about one thousand in number, provides a fascinating view into this kind of existence.[1] Providing further testimony are Kendrick Frazier, editor of *Skeptical Inquirer*, and his wife, Ruth, who visited Tanzania, where they met with some Hadza people and observed their hunting-and-gathering behavior.

The Hadza have existed by foraging for food. They never developed agriculture; there apparently was no need to, as food was plentiful enough to subsist on. They survived by hunting down and killing any animals they could capture: baboons, antelopes, warthogs, bush pigs, birds, zebras, and buffalo. They eat almost anything they can find and kill. The women gather berries, edible tubers, and baobab fruit. The men hunt animals and forage for honey.

As a result, the Hadza have few possessions—an axe, bow and arrows, knives, a pot to cook in, and a blanket. Since they move constantly in search of the hunt, they only build temporary shelters. Apparently there are few social structures, ceremonial rituals, or celebrations. There are no priests, shamans, or medicine men, no idea of what happens after you die. The Hadza share their food communally, and after the hunt, all join in the feast. Groups are small in number, approximately thirty people. There are few possessions, few social obligations, and apparently no religious structures. They have their own dialect by which they communicate. Women give birth in the bush by squatting. Half of all children do not survive to see their fifteenth year.

With the invention of agriculture, farmers grew and harvested crops and hence lived without the need to hunt. At this point, villages began to emerge, eventually cities, and then larger sociopolitical institutions, accompanied, of course, by the growth of culture.

The most important factor for humans is that social groups and cultural customs provide continuous protection for the young. If nature is hazardous for the lone individual, it is the family network, tribe, or community that offers care during the extended period of dependency.

Child psychologists tell us that although the brain is wired, providing instinctive guidance for the human being, it is capable of adapting to the environment in a multitude of ways. The child first learns the responses that are appropriate or inappropriate within the family/tribe/community for which it was rewarded, reprimanded, or punished. Obedience to social rules and customs—sociogenic constraints and latitudes—thus extends structures over and beyond instinctive biogenetic processes. For humans, the capacity for linguistic development and moral empathy has roots in their biological endowment; how these are developed depends on the nourishing care within the cultural context.

Thus education, learning, and conditioning offer a safety net for survival. The child does not have to hunt on its own, as does the leopard or wildcat, because it has been provided by the clan with the tools of sustenance and survival. The youngster is taught how to forage for food, how to gather and pick fruit, and is later given the know-how for hunting and fishing. Those tribes whose members domesticate animals and live off them learn to drink the milk or blood of cattle and use their skins for clothing and shelter. It is the community that grants protection at all stages of life.

The pivotal development in cultural evolution was the introduction of agriculture, the planting of seeds and reaping of crops. Here the clan could settle down, claim a piece of terrain or cave, build huts or homes, and protect itself for survival. Thus the perils of nature are overcome by the evolution of social systems and cultural traditions, and by the transmission of technological tools and skills, such as the scraper, knife, hatchet, or plow. Customary moral rules are also transmitted from generation to generation; they are products of group selection. The social group sustains the individual so that he or she is not left to fend alone; very few humans would survive in isolation. Culture is added to biology and is the chief impetus of human evolution in the past 11,700 years.

Sociocultural institutions thus provide the means to achieve a more peaceful, productive, and satisfying life. In time humans inhabited villages and cities remote from a rural setting, and a complex network of economic, political, and sociological institutions developed. Urban life shielded individuals from having to cope with the natural environment. The environment had become thoroughly sociocultural.

This catapults humans to a new level of complexity. New forms of competition, conflict, and aggression result, pitting social groups against each other in the struggle for survival. Personal responsibility for an individual's own future developed in the modern world. Autonomous choice provided new means for persons to survive and flourish. Today the starting point is not nature—Tarzan in the jungle brought up by chimpanzees or the legend of feral children brought up by wolves—rather, the original state for humankind has become culture, not simply nature by itself. In its most developed form, the arts and sciences provide the instruments to survive and thrive in highly complex social systems. Here rich cultures develop, offering linguistic, literary, artistic, religious, philosophical, and scientific learning; and thus civilization grows.

Does the uncertainty encountered in the natural world disappear entirely? Not at all. Though civilization brings the fruits of law and order, peace and tranquility, the healing arts, the luxuries of taste and comfort—and although humans are largely liberated from the brute forces of nature—there is still the turbulent struggle to achieve and succeed. Chance and indeterminacy are as prominent in culture as in nature, though they assume new forms and dimensions. It is not simply wild beasts or disease, droughts or raging storms out in the open that challenge human beings but also the clash of social systems and conflicting religious and ideological creeds. There are times of splendor and efflorescence, but there are also periods of war and devastation. In order to escape the vicissitudes of fortune, men and women have sought to create systems of laws and constitutions, which they hoped would endure through the ages. But these do not last indefinitely, and the erosion of time has reduced even the most impregnable of social institutions to naught. Political absolutes erected to stave off change are invariably undermined by new dangers and incursions, and the challenges of new ideas and ideals are ever present. We can see these dramatic processes at work in the emergence and decline of civilizations in the past and the present: history is the best evidence for the persistence of turbulence in human affairs. It is the oracle for what will happen in the future.

THE RISE AND FALL OF CIVILIZATIONS

Arnold J. Toynbee, the British historian, has written extensively about the rise and fall of civilizations.[2] By *civilization* he was referring to a complex "cultural and religious outlook" that dominated an age or region and eventually declined or was supplanted. Among the many civilizations that we can discern in history were the Sumerian, Egyptian, Babylonian, Assyrian, Judaic, Greco-Roman, Persian, Indian, Chinese, Japanese, Islamic, South and Central American, and Western civilizations. He attributed their fall to whether they were able to respond to the challenges that confronted them. Although there are periods of peace and tranquility, nothing in human affairs is fixed forever, and turbulence has been a constant feature of every civilization. The only thing that does not change is change itself. The barbarians are always at the gates. New forces within are ever ready to pounce, seize power, and overthrow the ancient ways.

Will and Ariel Durant's impressive eleven-volume work titled *The Story of Civilization* gives an overview of civilizations from the Oriental heritage and classical Greece and Rome, through the medieval period, on to European culture and the modern era.[3] The Durants are truly comprehensive in what they seek to accomplish, providing chronicles of the various histories of economics, politics, religion, science, philosophy, music, and the arts. Will Durant in the early volumes, and then with his coauthor, Ariel, presents richly detailed and unified accounts of historical periods. Beginning in the Orient they trace various civilizations down to Europe in the nineteenth century. The Durants make it clear that these histories reveal tempestuous times, full of brilliance and creativity, interwoven in complex civilizations yet eventually engulfed by uncertainties and contingencies in every epoch. There are clearly analogies, similar patterns, familiar motives and passions displayed by real human beings. There are failures, defeats, disasters, and tragedies, as well as remarkable discoveries and achievements, times of war, but also times of relative harmony, peace, and prosperity.

Throughout the saga is the evolution of economic activities, from hunting and gathering to agriculture and tillage, barter and trade, craftsmanship and industry, invention and technology. Political elements are apparent: from the clan and tribe, to the village and city, and to the growth of the national state and empire. Similarly for power and control and the conflicts between contending factions—there are in every civilization periods of strife and conflict as well as times of relative tranquility. They depict the role of military force and the evolution of weaponry, tactics, and strategy in conquest and defeat, but they also provide accounts of how morality evolves. These periods describe the wide diversity of moral codes in history, but they also point out the common ethical principles that are shared—from the Code of Hammurabi, the Ten Commandments, and the Sermon on the Mount to modern principles of ethical humanism. There is a plenitude of gods and goddesses, from Isis and Osiris in Egypt to Zeus and Apollo on Mount Olympus. There are bizarre rituals, which vary from region to region—from the role of animistic magic and the superstition of primitive folk to the emergence of the tales and parables of Christianity, Islam, Judaism, Buddhism, Hinduism, Confucianism, and Shinto. Different civilizations adopt diverse modes of dress and costume, manners and mores, from Babylon to China, Africa to Scandinavia, India to Persia; similarly for the great diversity in dialects, languages, oral and sign languages, and written modes of communication.

Diversity in human culture is seen in the different kinds of foods prepared and consumed—from raw fish and whale blubber to horse meat, from geese and birds to snails and insects, nuts and fruit. And there is a plethora of fermented beverages from wine and beer, sake and tequila, cognac and scotch; and also staples: bread and potatoes, rice and noodles, beans and manioc. Delicacies include Roquefort® and marzipan, caramel and chocolate, mint and taffy!

The modes of relationships are diverse, from monogamy to bigamy, exogamy, polygamy, and polyandry, harems, group marriages, trial marriages, open marriages, prostitution, romantic love, chastity, celibacy, patriarchy, and matriarchy. There are various customs governing the bringing up of children. There are different legal systems for settling disputes, dealing with revenge, recompense, and accountability: the law of retaliation (lex talionis), the substitution of damages for revenge, Roman law, the Magna Carta, the Napoleonic Code, the Rights of Man, the Bill of Rights, and the Universal Doctrine of Human Rights. The splendid works of architecture of past civilizations are today the common heritage of humanity. We can appreciate and enjoy their diverse forms and shapes, colors and hues: the massive pyramids, temples, mosques, and arching cathedrals for worship; the statues of Michelangelo and Rodin; and the splendid paintings—the prehistoric drawings in the caves of France and Spain; the depictions of Christian and Muslim mystics; the exotic works of art of China, India, and Africa; and the paintings of Rembrandt, Renoir, and Picasso. I recently visited the Musée Matisse in Nice, the former mansion where Henri Matisse lived. Next to it was an outdoor archaeological museum of a Roman amphitheater where many public events were carried on—though the stone arena was decaying, the stark elegance of Roman civilization still displays its once-great grandeur.

We can learn much from the great philosophers of the past—Aristotle, Spinoza, Kant, Hume, Derrida, Russell, and Peirce, and from scientific discoveries that advance the frontiers of human knowledge—from Newton and Galileo to Darwin and Einstein, Watson and Crick. The study of the rich multiplicity of civilizations enables people in any one civilization to overcome chauvinism and insularity. An appreciation of the civilizations that humanity has created opens us up to new horizons of imagination and ingenuity.

Will Durant thought that the conflict between magic and myth versus science and reason is ongoing in human history. He observed that the priestly control of science and the arts both shackles the mind and impedes the improved well-being of humankind; that as knowledge grows, anticlericalism leads to the decline of religious superstition. The former theological unity of belief and morality in a civilization may eventually be replaced by a return to reason. But this, in time, may degenerate into disillusionment and moral hedonism unless there are viable moral principles to replace those that are discredited. Alas, in time these, too, may be replaced by new mythologies. The range of world civilizations is so vast that I will only focus with broad brushstrokes on a few to depict the general traits of cultures and civilizations—as revealed by the writings in books, the arts, architecture, and monuments—testifying to what they meant to the people who lived in them.

GREEK AND ROMAN CIVILIZATIONS

Let us begin with Greece and Rome. Students of the classics are familiar with the Golden Age of Hellenistic culture, the fountainhead of Western civilization. The grandeur of Athens comes readily to mind, with its creative achievements in philosophy, logic, science, literature, the arts, architecture, and the law; but also well-known are its turbulent military, political, and economic conflicts. We remain profoundly influenced by its authors, philosophers, poets, artists, and architects. We read the Socratic *Dialogues* of Plato with profit; the brilliant scientific and philosophical works of Aristotle; the plays of Aristophanes, Euripides, Sophocles, and Aeschylus; and we marvel at the stately beauty of the Parthenon and other classical monuments and statues.

Broadly viewed, ancient Greek civilization grew out of the Minoan and Mycenaean cultures, the latter of which collapsed about 1150 BCE. The origins of this civilization can be seen on the island of Crete and through the epic poems of Homer, the *Iliad* and the *Odyssey*. In Homer's heroic tale, the Trojan War was provoked when Paris of Troy (in Turkey) eloped with Helen, wife of Menelaus. The Greeks, led by Agamemnon, launched a thousand ships to storm the city and recover her. Among the great Greek warriors was Achilles, who died in battle.

Greece began to emerge as a great commercial and seafaring culture by the eighth century BCE. Among the leading city-states were Athens, Sparta, Corinth, and Thebes. The Olympic Games were first played at Delphi in 776 BCE and offered a truce between warring city-states. Despite almost continuous wars among the Greek city-states, interrupted only by the Persian wars (499–479 BCE), Greek culture flourished. Greek settlements were spread beyond the Aegean and Ionian Seas, and colonies were established in southern Italy, Sicily, Asia Minor, and throughout the Mediterranean world.

Philip II of Macedonia conquered the mainland of Greece. His son, Alexander the Great, a pupil of Aristotle, raised an army to defeat the Persians. He conquered the city-states of Greece and swept on through Asia Minor into Egypt, Asia, and Persia, over the Hindu Kush to India. He thus extended the influence of Greece far and wide. The influence of this brazen, heroic young man inspired his army to follow him to the far ends of the world, pushing ever deeper into Asia Minor, bringing the ideals and beauty of Hellenistic culture to distant places. Alas, the peak of Greek power ended in 323 BCE when Alexander died unexpectedly, though it was not until 146 BCE, when Greece was conquered by Rome, that Greece's political independence ended, and even then the influence of Hellenistic culture on the Roman Empire continued unabated. The Stoic, Skeptic, and Epicurean schools of philosophy that flourished in Rome were Greek in origin.

Hellenistic culture spread throughout the Mediterranean world, though it was constantly embroiled in conflict. The observations of two great Greek historians, Herodotus and Thucydides, provide invaluable accounts of warfare between Athens and its neighboring city-states and led to conjecture that there were perennial traits of human nature that precipitate turbulence and dissonance in human affairs.

Herodotus, who lived in the fifth century BCE, wrote a nine-volume work describing the Greek-Persian wars, including the ultimate invasion and defeat of the Persian army in 480 and 490 BCE. He depicted the legendary heroic stand of a small number of Spartan warriors at Thermopylae. Thucydides gave an engrossing account of the later Peloponnesian war between Sparta and Athens, of twenty-seven-year duration near the end of the fifth century BCE. This is reminiscent of the rivalry between England and France in modern times or, later, between Germany and France. His astute analysis attempted to be objective, describing the wars between city-states in which questions of military might trumped considerations of ethical rectitude and morality. Thucydides gave graphic accounts of power politics as it was played out in wartime. Being an officer in the early part of the war himself, he was exiled by his Athenian compatriots from 424 to 404 for losing a battle in the first years of the war. His trenchant observations of human nature and how wars permit the worst tendencies of human beings to commit atrocities are appreciated today for their insight. This was a classical commentary on the human condition. The reflective observations of both of the above historians say something perennial about human nature and how it can turn mendacious under warlike conditions, allowing individuals to act with extreme hostility toward others. Especially eloquent was Thucydides's account of the great speeches delivered by Athenian political leaders, such as Pericles's funeral oration, in which he provided a defense of Athenian democracy, contrasting it with the Spartan military state. These speeches are often fictitious, serving as a vehicle for Thucydides to present his own reflections.

The question that has been raised concerns the role of philosophical, ethical, and religious factors in taming the ferocious beast within the breast of men—which when unleashed knows no bounds of forbearance. These tendencies have been evident as far back as can be traced in human history. Whenever all-out warfare replaces normal civilized conduct, the concern for the rights and duties that human beings owe to other humans seem to disappear. What is apparent to any dispassionate observer of human behavior is that when men and women are released from social constraints (customs, laws, religious or moral codes), they often behave like wild beasts. They are at such times wont to plunder, rape, torture, and murder those they consider their rivals; they are no longer humane in any meaning of that term; their enemies needed to be defeated and even destroyed.

The view of man as selfish is outlined by Thrasymachus in Book I of Plato's *Republic*. He illustrates this with the myth of the Ring of Gyges, which enables a person who rubs it to become invisible: would man commit foul deeds if he were immune from social disapprobation? Man, asserts Thrasymachus, is by nature nasty, brutish, and self-seeking. This view was later echoed by Thomas Hobbes, who believed (mistakenly, I think) that men are by nature selfish. Thrasymachus says that if someone were assured of being undetected, he would murder the king, rape the queen, and take possession of the kingdom. Socrates disagreed with this view of human nature; he proceeds in the rest of *The Republic* to elaborate an ethical theory that allows the ideas of the good, justice, and beauty to serve as beacons for the life of reason and justice.

The primal impulses in human nature—especially the instinct for aggression—can be traced back to prehistoric origins. Many have attributed these primarily to the male, for he is physically stronger than the female. He is no doubt stimulated by raging testosterone, which predisposes him to commit warlike acts. This is particularly the case when there is a breakdown of moral standards, as often arises between contending tribes, clans, cities, or states. It is the males primarily who assemble armies in order to vanquish those perceived to be their intractable foes.

Thus a potential impulse toward violence can be triggered whenever there is a breakdown of moral constraints. This impulse is surely contagious, and females may share in the fear and hatred of their adversaries, and so they may join their mates in goading them on to battle, cheering them when they go forth, praising them when victorious, consoling them when wounded, and mourning those who have fallen—even though females do not have the same hormones to contend with and are as a rule more pacific. The male thrust in sexual intercourse is active and aggressive; his erotic passions are aroused when he can ejaculate. The woman is more passive, receiving the sperm with erotic enjoyment and orgasms that can be experienced many times.

It is of course not possible to draw a sharp line between male and female, for humans are androgynous (on a bell curve), having both masculine and feminine characteristics. Men and women may be sexually aroused by a wide range of fantasies: submission and dominance vary from individual to individual, and some women may be as aggressive and self-affirming as men, while others seem to prefer the passive role and enjoy submitting to a dominant partner, male or female. Many men crave ambition and power over others, but surely not all.

In any case there are many impulses and passions that impel men and women to various forms of behavior, not the least of which is the moderating role of reason. The Greeks observed that humans are powerfully motivated by their emotions, and if they are intense, they may be overcome by them. On the other hand, they are capable of reflective considerations; rational deliberations may induce them to restrain their passions, to be temperate in their cravings. Social conditioning may also train children not to give in to their capricious temptations and to resist every wild wish or desire. Lust, greed, avarice, power, and sloth need not dominate a person's life. A person may wish to disrobe in public or pursue anyone who arouses his or her libido, yet social disapprobation prevents him from giving in to any desire that arises. A mature person recognizes that reason and prudence should arbitrate between conflicting passions and balance them.

There is still another factor that is often overlooked, and that is the importance of same-sex bonding, which may, and indeed often does, exert a powerful role in human conduct. It is most clearly seen on the battlefield, where men soon discover that the only way to defeat the enemy legions is by united action among the men on their side against their opponents. This is sometimes described as "team spirit," and it is perhaps best seen in sports contests, where everyone recognizes that if a soccer or football team is to achieve victory over competing teams, it needs to play in unison. As cheerleaders on the sidelines arouse spectators to goad the home team, so the team players must bond with their teammates. Sports teams are perhaps the moral equivalent of actual warfare, and the deep impulse to defeat the enemy in mortal combat, until one or the other is killed or wounded, is played out in dramatic and evocative form without the desire to maim players on the opposing team.

We can speculate about the roots of male bonding, but they may be psychosexual in part. The Greeks recognized this, for they adored the nude male figure and erected statues in the public square to express this adulation. A well-known Greek saying was that the most effective army was made up of lovers who fought together side by side in a life-and-death struggle to vanquish the enemy and survive. This was surely the case with Alexander the Great. Always beside him was his boyhood friend and lover, Hephaestion. Male love did not inhibit Alexander to the lure of beautiful women. He encouraged his troops to marry the women of the countries they conquered, and he himself married Roxanne, a Bactrian princess, and sired a child with her. He also wed Barsine, the oldest daughter of Darius, and others. He most likely was bisexual.

Greek culture thus extolled the love of males for other males, particularly the affection between a youth and an older man, whose task was to provide moral guidance. The great Greek poet Homer, in *The Iliad*, dramatizes the deep affection that Achilles had for his young lover, Patroclus. The agonizing death of Patroclus is especially moving as Achilles returns to the scene of battle to avenge his death. Did this bond imply sexual pederasty or very strong friendship between comrades-in-arms? The Greeks preferred the company of men for pleasure and conversation and women to make babies. The Greek poet Hipponas observed: "The woman is twice a pleasure to man, the wedding night and her funeral."

Now, surely male bonding or indeed similar affection of women for each other does not necessarily indicate a gay or lesbian relationship. Yet same-sex bonding is a vital relationship between men or women to keep the social community together. The polarity for males is that both aggression and affection (friendship) are present in human relationships.

Greek hegemony in the Mediterranean region in time was supplanted by Rome, which became the predominant military and political power of its day, even though it revered Athens and borrowed heavily from Hellenistic culture. Much has been written about the magnificent rise and eventual decline of the Roman Empire, which dominated much of the European, Near Eastern, and Mediterranean regions. This was based on military might and law, and it produced its own literature, which we read with profit. Roman citizenship was prized, and the empire provided excellent roads as a basis for commerce and trade, even though it rested heavily on the backs of conquered nations and slaves.

The Roman Empire was built and maintained by its legions, who were all too willing to engage in bloodletting to maintain law and order. The noble families during the Republic, or the emperors, would torture and kill their opponents to maintain power or exact tribute. A good illustration is the fate that befell Carthage, which had become a great commercial city in Tunisia on the North African coast. Carthage began to rival Roman hegemony. Hannibal conquered Spain in 218 BCE; he invaded Italy through the Alps while carried on the backs of elephants—a daring military exploit indeed! He left a path of destruction through Roman cities and a huge toll of dead Romans. After decades of warfare between Rome and Carthage in the Third Punic War (149–146 BCE), Rome ordered Carthage to be completely destroyed. It was invaded, its men were slaughtered, its women and children were sold into slavery; no building was allowed to stand, and no stone was left unturned. The very existence of Carthage was extinguished in 146 BCE. A similar fate—not as drastic—was later to befall Rome, as the Western Empire was overwhelmed by barbarians from the north and east, including Attila the Hun.

Turbulence, slaughter, and destruction are widespread in human history, and literally tens of thousands of wars, uprisings, revolts, revolutions, and civil wars have been waged in human history, and these appear to be an enduring feature of human civilization. The whole thrust of human affairs is to quell the ever-threatening barbarians at the gates—and this it is often able to do by waging violent wars. Alas, no civilization thus far has withstood the buffeting of time, and even the greatest have declined, though hopefully the lessons of our violent past will enable humans to overcome the military option.

The noble humanistic ideals of pagan Greece and Rome were eventually overthrown by barbarians, and the mystery religion of Christianity spawned in the Middle East engulfed Rome. And this in turn was challenged by Islam, which swept everything in its path, from the Atlantic to India. Medieval Christianity was replaced by modernism, and this is now challenged by virulent forms of fundamentalism. There is no guarantee that a rich, creative, humanistic culture will not be inundated by new forms of mythology in the future. There is no assurance that progress will continue, or that the absolutists in some distant epoch will not destroy all in their way, or that reason will not again give way to new forms of spiritual madness. Religious passions may exert destructive or inspiring influences that may in time develop into a new civilization, as did the rise of Islam in the seventh century. Nevertheless, human beings can learn from history, and there is the strong desire to create new planetary institutions and a system of world law that will enable societies of the future to settle their differences peacefully.

THE EXTINCTION OF CIVILIZATIONS

History teaches us that the civilizations of the past, like all other creations in human history, are mortal. We all know that but often seem to forget it. The list of civilizations that came into being, developed at first with spurts of intensity and hope, flowered in brilliance and splendor, then began to decline with pessimism and exhaustion, and eventually were overrun and disappeared, is extensive. This process is nigh universal, indeed endemic to human civilizations. But can it be overcome in the future?

Permit me to focus on civilizations in South America that have declined, notably the Mayan civilization. I do so because lost civilizations in Europe are usually replaced by others, such as Greece or Rome in the Mediterranean or the Khazar kingdom in Russia. But this was not true in South America, where the Mayan civilization virtually disappeared, its cities overgrown by jungle. Israel is a special case; it is an anomaly because the ancient Hebrew civilization of the Jews, which was located in Palestine, was decimated by the Romans and barbarians and would have totally disappeared had it not been for the Bible, which kept Israel alive in the hearts and minds of the Jews of the Diaspora. Israel was reestablished in 1948 as a Jewish homeland after the Nazi Holocaust had decimated Jewish settlements in Europe and forced many Jews to long for a return to Palestine and reestablishment of the ancient state of Israel. This was fed by religious feelings that God had bequeathed Israel to the Jews, though a large number of Zionists who established Israel were thoroughly secular.

The Mayan civilization of the Yucatán Peninsula of Mexico, Guatemala, and Honduras stands high on the list of extinct civilizations because it had achieved its own greatness, and, like others, it considered itself immortal. Yet it is no more, though we can imagine its former impressive achievements by examining the remnants of its once-great cities and pyramids in the heart of the jungles of Central America, now overgrown and uninhabited. These ruins attest to the creativity of its people, the architecture of its proud constructions; its monuments and sculptures attest to its former confidence without limits of its destiny.

The apogee of the Mayan civilization, the most advanced in the Americas, was about 800 CE. Its royal rulers adorned themselves with colorful costumes; they sacrificed young girls to placate the gods (they did so by tearing out their hearts). Its great monuments and stone cities have long since been abandoned.

There are remnants of other civilizations that can also be seen in South and Central America, such as the Nazca civilization of Peru, 240 miles south of Lima. Many today are familiar with the huge drawings etched on the Nazca plains in Peru. Archaeologists believe that this civilization existed from 500 BCE until 650 CE.

Another civilization was Tiwanaku in Bolivia, southern Peru, and Chile. Here we can locate the pyramids of Akapana. This civilization apparently lasted from 1500 BCE to 1200 CE, with its apogee in the tenth century. The Tiwanaku civilization discovered the use of bronze, which gave it a great advantage over other peoples. The environment, particularly the lack of water except for rainwater, was most likely the contributing factor in its decline. All of this occurred before the arrival of Western European invaders centuries later. It dramatizes the stark fact that no civilization is eternal. Generally they have endured from two hundred to six hundred years. Many civilizations are destroyed by cataclysmic events, such as an epidemic or plague (the Black Death in Europe wiped out large segments of European populations). Other causes may be environmental, such as the volcanic eruption that destroyed Pompeii or a great flood or earthquake. Of course, the destruction of once-proud civilizations has often been caused by invading armies. This means that human beings are the main reason why many civilizations have disappeared.

The ruins of the Altun Ha ceremonial site in Belize, Central America, are a stark reminder of the once-tremendous prominence and subsequent decay of the Mayan civilization that constructed it. The remains of thirteen structures (some pyramid-like in form) and three plazas are evidence of a developed culture and agricultural economy. The Mayans grew corn and avocado fruit and ate wild boar for meat. Their Altun Ha site was the epicenter of a thriving trade that gained power and prominence in the Mayan classic period (250–99 CE). It was situated in a pristine rainforest that provided sufficient food and water and supported an estimated population at its peak of one million inhabitants. Unfortunately, it rapidly declined due to environmental deficits. There was constant warfare and battles with rivals that no doubt took their toll. But the lack of water apparently was chiefly responsible for the civilization's collapse. The Mayans depended on rainwater that they collected in basins. There were few running streams or deep wells, and so with warming periods and droughts they were unable to support the once-flourishing population.

My personal visit to the area in 2009 surprisingly indicated that remnants of their once-proud religion (it worshipped Sun, wind, and rain gods) still persisted; at least some of its ancient traditions are still being practiced. Many descendents of the Mayans still observe "the Day of the Dead." Once a year they light candles and put out an offering of food for their deceased relatives and friends. This ancient rite has been incorporated into Christianity, which the Spanish conquistadores and priests brought to the native Indians in the sixteenth century, long after the collapse of the Mayan kingdom that was subsequently overgrown by the encroaching rainforests of Central America, which have erased most of the ruins of this once-proud civilization. The Mayans are noted for their astronomical observations, their use of mathematics, and their own ingenious calendar, which many moderns still find intriguing, especially its perceived prophecy of an end-of-the-world scenario.

Scientists and historians have attempted to explain the rise and fall of civilizations. For example, Jared Diamond, a biogeographer and evolutionary biologist, asks why human history unfolded differently on different continents over the past 13,000 years. Diamond is especially concerned with responding to the question of why Eurasians (people of Europe and Eastern Asia) expanded so rapidly around the world, especially after 1500. He maintains that any scientific account of historical change should be in terms of natural and material causes. His interdisciplinary research project draws on biogeography, archaeology, linguistics, molecular biology, and plant and animal genetics. He asks: Why did the indigenous peoples in Africa and Australia and the native Indians in North and South America (the Mayan, Inca, and Aztec civilizations) so easily collapse?[4]

Diamond maintains that it was because the people of Africa, Australia, and

Latin America remained with Stone Age technology as hunter-gatherers, whereas Eurasians developed agriculture, metallurgical technology, and political organizations that outmatched the underdeveloped societies they conquered. More specifically, he claims that guns, steel, plant species, domesticated animals (horses and cattle), and oceangoing vessels led to the collapse of backward cultures. Of special significance is the fact that the indigenous Indians of the Americas were not immune to the germs transmitted from Europe, and as a result millions were wiped out by infectious diseases. For Diamond, *geography* is a key factor in the supremacy of a civilization: its location, climate, natural resources (fresh water, fertile soil, potential food supply, minerals), and whether it is located in coastal regions for civilizations that engage in commerce and trade.[5]

Karl Marx proposed the sociological interpretation of history to explain social development. He maintained that the economic forces and relationships of production were the basic causal factors in social change, and that the level of sociocultural development was a result of these economic causes. There are different modes of production: agricultural (the harvesting of natural resources and crops); manufacturing, commercial, or scientific-technological in the knowledge industry; and there are different sources of labor power: domesticated animals, human labor (slave, serfs, workers), or technical know-how. There are also different sources of energy: the combustion of wood and coal, oil, and gas; as well as steam power; electricity; nuclear power; and more. Marx considered the relationships of production; that is, the class structure of the social system as part of the economic base—emperors, nobles, citizens, and slaves in the Roman period; landowners and serfs during feudal times; the bourgeoisie and craftsmen in mercantilist societies; and the proletarian workers and capitalists in modern industrial and commercial economies.

According to this interpretation, the political, religious, moral, intellectual, and cultural characteristics of a society are part of the "superstructure" and are a result of the economic base. The economic fundamentals include the key role of the relationships of production—the way society is organized to produce and distribute goods and services. Marx also postulated the surplus theory of value in the form of wages, rent, interest, or profits. He said that although human labor power provides value, it was not equitably distributed. This class analysis overstated the role of classes in causality, for factors in the so-called superstructure—political, religious, and scientific—may at times supersede economic causes. Nevertheless, the Marxist emphasis on economic causes is a powerful insight in explaining societies of the past. It does not, however, adequately account for information societies that have emerged where new modes of production have appeared, such as individual contractors working in remote locations on computers.

Actually there are many causal factors at work in the rise or decline of societies or civilizations. The role of contingent events in history is often decisive; many of these may be unforeseen or accidental. Similarly for the role of great individuals or heroes in history (Napoleon or Jefferson), great religious prophets (St. Paul or Mohammed), or moral leaders (Gandhi or Martin Luther King). Innovative technological discoveries may rapidly transform a society—for example, the invention of electricity or antibiotics, automobiles or locomotives—and these may have unforeseen consequences. Let us leap ahead to the contemporary world to illustrate the perennial appearance of turbulence in human affairs.

WARS IN THE TWENTIETH CENTURY

There are many motives in human nature and socio-economic-political causes that are responsible for conflicts, revolutions, and wars. These include aggression, jealousy, greed, lust for power, hatred, the penchant for plunder and conquest, national pride, racial and ethnic hatred, and economic competition for markets. There are also misguided hopes for victory, fear of the enemy, and the passion for glory. The role of contingent events in human history is of course often decisive. Thus, the assassination of the archduke of Serbia, Franz Ferdinand of Austria (heir to the Austro-Hungarian throne), and his wife, Sophia, in 1914 in Sarajevo, capital of Bosnia and Herzegovina, ignited World War I, which began a month later. The assassins were nationalistic Serbs who wished Serbia to break away and establish an independent country, which they did after World War I, as part of Yugoslavia. Another contingent event, the bombing of Pearl Harbor on December 7, 1941, by the Japanese, led the United States to declare war on Japan.

Of course both of the above historical events—the assassination and the bombing—were part of already-inflamed political and economic conditions, and there were trends that were at work. In World War I, there were economic, nationalistic, and military rivalries between the great powers of Europe—France, Britain, and Russia versus Germany and the Austro-Hungarian Empire. In World War II, Germany and Italy had allied with Japan against the United States and Britain, and although the Japanese attack was the specific event that led the United States to enter the war, there was no doubt that underlying economic, political, ethnic, and military factors helped to bring it about. The Japanese claimed that trade barriers prevented them from free trade, and they tried to cripple the United States for its economic strangulation of Japan. Yet these events did occur, and without them subsequent events may have been different.

Another cause of social change is *ideological*. Prior to World War II there were profound differences in ideas and values between Nazi Germany and the West. Fascism was a nationalistic movement that condemned the democracies of the Western powers and extolled the heroic virtues of *Macht*. Mussolini emphasized the power of the fascist state and attempted to restore the martial virtues of the Roman Empire. Hitler frightened the democratic countries and caused them to rearm; this fear was exacerbated as Nazi Panzer tanks and the Luftwaffe provoked a domino effect, and country after country fell. Hitler blamed the Treaty of Versailles after World War I as the reason for his grievance. He said that Germany needed *lebensraum* (living space); Western colonial empires dominated the world, and Germany was left out in the cold. He affirmed that the blond, blue-eyed Aryan race was superior to all others, and he wished to establish a New Order throughout Europe. He sought to dominate or exterminate those he considered inferior, such as Poles, Slavs, and Jews. Hitler built his ideology on anti-Semitism, attributing Germany's plight to a Jewish conspiracy. He condemned the Jews for being both finance capitalists and communist commissars. The magnetic personality of this former Austrian paperhanger inflamed the German public, whose economy was in tatters, and who were angered by the Treaty of Versailles. It was the impassioned oratory of Hitler that was able to rouse them to support rearmament in preparation for World War II. Although many were opposed to him, there was little they could do to confront the fascist dictatorship that Hitler imposed. The role of mendacious leaders in history has led to untold bloodshed and suffering.

I served in the US Army during World War II and was in the army of occupation, stationed in Munich, Germany. What fascinated me were my many nocturnal visits to the Munich beer hall where Hitler first got his start. He was imprisoned, eventually released, and went on to build the National Socialist (Nazi) Party and to defeat the left-wing opposition of the communist and social democratic parties. How Hitler managed to seize power and persuade the German population to support him is a lesson in political deception and power. Germany was a great nation in which science and philosophy, music and metaphysics flourished; nevertheless, it succumbed to a madman. We should be grateful that Germany has since recovered after World War II—itself a kind of miracle, that it eliminated the Prussian Junker military class that dominated Germany so long, and that it was able to build a democratic society and join other democracies, full of contrition about the Nazi episode. It was willing to take part in building a new European Parliament and the European Common Market by eschewing the rabid nationalism of the past. Germany thus redeemed itself in the eyes of the democratic world—in spite of the horrors of its Nazi past. This demonstrates how the tumultuous gyrations that can traumatize an otherwise civilized nation can be overcome.

Adolf Hitler was a demonic personality who seized power in 1933 and began to rearm Germany and prepare for another war in which he could redeem the greatness of the Fatherland. Especially intriguing was his rise to power and his later hypnotic influence on a substantial portion of the German people. (I say this because of personal reasons. I enlisted in the US Army in 1944 at the age of seventeen and a half. After only six months of basic training, I was dispatched to England in 1944 and then hastily rushed to Europe during the Battle of the Bulge, where Hitler made a last effort to split the Allies by sending the Wehrmacht in attack against the Western Front. I was with Patton's Third Army, which subdued Germany, pushed all the way on half-tracks to Pilsen in the Sudetenland. We came back to Munich, where I was stationed. It was a city virtually destroyed.) The ideology of the Western countries countered fascism and defended democratic values. Indeed, Woodrow Wilson declared that World War I was waged "to make the world safe for democracy." Thus this new international confrontation was manyfold—political, economic, and military—but it also contrasted profoundly different ideological perspectives. The democrats believed that all humans are of equal dignity and worth, and they sought to defend human rights. The fascists adamantly rejected these ethical principles.

Actually, many historians have maintained that on the eve of World War II Franklin Delano Roosevelt was convinced that the United States needed to enter the war and that he began providing arms to Britain—such as the agreement to lend/lease fifty destroyers. FDR was a consummate politician who was able to rally the American public to support the defense of democracy against the rising fascist tide. He was a patrician who was hated by the ruling classes, yet he was able to succeed in forging a powerful alliance against the growing power of Germany. America felt a strong affinity to Britain, and so Pearl Harbor was used as a rallying cry to overcome strong isolationist voices in America. Even though many democratic theorists have been critical of the resort to violence to resolve differences between nation-states, many thought that the war against fascism was a just war. Earlier, many idealists joined the left wing "Abraham Lincoln Brigade" against Francisco Franco's fascism during the Spanish Civil War. This ideology is not simply a reflection of the economic structure of a society but is itself a causative factor. The motives and intentions of human beings enter into history and change things fundamentally. Ideas have *consequences*.

One ideological complication that arose in the 1930s was the specter of totalitarian communism. Many in the West, even in Germany before the rise to power of the Nazis, were sympathetic to socialism, though the term was interpreted in different ways. There were many political variations on the Left, and there were often intense disagreements between democratic socialists on the one side and communists, especially of the Leninist-Stalinist variety in the Soviet Union, on the other. The communists had explicitly defended the use of terror. Any means can be taken, they insisted, to seize power, and once in power to consolidate it. There was an intense battle between the Mensheviks in the Russian Duma—they were social democrats—who believed in the electoral process, and the Bolsheviks, who maintained that *any* type of violence must be used during a revolution to wrest control from the czarist regime in Russia and to overthrow capitalists in other countries.

The Western countries of France and Germany had strong socialist and social democratic parties before the First World War, with leaders such as Jean Jaurès of the French socialist movement and Leon Blum in the 1930s; and in

Germany Karl Kautsky and Eduard Bernstein. The defection of Leon Trotsky from the Soviet Union—in his dispute with Joseph Stalin—had considerable influence among Western democratic socialists, for Trotsky had exposed the mass terror and purge trails of Stalin. Trotsky was murdered in Mexico City in 1940 on orders from Stalin. Trotsky was one of the top Bolshevik leaders of the revolution and founder of the Red Army. This controversy split Western socialists—many on the Left refused to believe it; others did and were so repelled that they considered Stalin a monster equal to Hitler. In Britain, George Orwell lampooned Big Brother in *1984*. In the United States, Max Eastman, Sidney Hook, John Dewey, and Norman Thomas defended democracy and attacked totalitarianism. In Europe, Rosa Luxemburg and Arthur Koestler waged a strong campaign against Soviet communism. A mock trial, convened by John Dewey and Sidney Hook, among others, before the outbreak of war concluded that Trotsky was innocent of Stalinist charges that he had betrayed the Soviet Union. But they rejected his advocacy of terror, and they were stalwart in their defense of democracy.

Thus on the eve of World War II many democrats were opposed to both Hitler and Stalin. It was the Nazi-Soviet pact in 1939 that summoned democratic socialists and liberals to unify in their support of the war and in opposition to fascism. It was only after Hitler broke with Stalin in 1941 and invaded the Soviet Union that diehard communists turned against Germany and joined in the war against Nazism. I recite all of this since I lived through it and vividly remember the intense controversies on the Left among liberals. Conservatives hated both socialism and communism with a vengeance. Although libertarianism did not have an impact until after the war, when economists such as Ludwig Von Mises and Friedrich A. Hayek, as well as social theorists like Ayn Rand and others, argued for free markets and against government interference. But I am getting ahead of my analysis of the role of contingency in human history and how it is often unpredictable. Stalin betrayed the communist revolution. His resort to terror was fed by his megalomania; and Hitler's anti-Semitism virtually brought Europe to its knees. Had neither men seized power, historical events most likely would have moved in other directions. Thus the role of malevolent individuals in history cannot be overstated. But of course there are other causes for the sudden and unexpected in the unfolding drama of history.

I have lived long enough to have witnessed the rise and fall of empires and civilizations or at least the "residue" of what remained—the Ottoman Empire (led by the Turks), for one, which held power for over five hundred years and was defeated in World War I. In visiting Eastern Europe and the Middle East today, we can still see remnants of a once-great empire, which included Greece, Bosnia, and Serbia in its grip, until forced to withdraw, and also its former far-flung control of North Africa and the Middle East. Similarly for the Austro-Hungarian Habsburg Empire, which was a major land power until its defeat in the First World War. Visiting Vienna, still a magnificent city, is instructive. Once built for an empire, the center of science, philosophy, music, and the arts, the remains of this capital city, which commanded a large domain, is today an empty shell, devoid of its great might. Before World War II the Austrian-Hungarian Empire was home to psychoanalysis and logical positivism, and it cultivated great scientists and philosophers (Freud, Wittgenstein, etc.). It supported the great symphonic works of Mozart, Beethoven, and Brahms. It flourished with luminescent flare and grandeur. But it collapsed. Today, Vienna is only the capital of the tiny country of Austria.

Barbaric genocides have occurred throughout the twentieth century. Witness the massacre of 1.5 million Armenians by the Turks of the Ottoman Empire during and after the First World War, including the forced marches, deportation, and rapes of its women. The murder of eight hundred thousand to one million Tutsis by Hutus in Rwanda in 1994 is another terrible spectacle. The Western democracies stood by and failed to act. The Nazi Holocaust during World War II, which slaughtered and incinerated millions of Jews, Gypsies, homosexuals, and the mentally handicapped, defies human comprehension—all done in the name of a theory of racial superiority.

Genocide of course can be found throughout human history. Often overlooked by the democratic West is the forceful capture of Africans and their transportation in filthy slave ships across the Atlantic Ocean (large numbers died on the way to the Americas) and their cruel fate after being pressed into slavery by their owners (with Bible and cross in hand). Similarly Native Americans were vanquished, and their cultures were virtually destroyed by European colonialists who considered them savages. If we go further back over the previous centuries we can view with dismay the murder and mayhem perpetrated in the name of God, such as the Christian crusaders on the way to Jerusalem slaughtering Muslims and Jews, or the ruthless killing of nonbelievers by defenders of Islam, who still today are all too eager to kill and torture both non-Muslims and those Muslims who disobey sharia's religious dictates.

In the twentieth century the First World War literally destroyed the flower of youth of Germany, France, and Great Britain. The Second World War killed tens of millions of people in bombed-out cities from Russia and Germany to England and France, from the Philippines to Japan. This culminated in the unleashing of two atomic bombs on the Japanese cities of Hiroshima and Nagasaki, where almost two hundred thousand innocent men, women, and children were incinerated or maimed. What is this, if not genocide?

Violence carried out in the name of God and salvation, fatherland and the flag, or democracy, peace, and prosperity has been endemic in human civilizations.

These phenomena were not confined to the Western world but occurred in Asia as well, from China and India to Korea and Japan. The history of China details a protracted series of wars, from the building of the Great Wall of China to keep out invading hordes to more modern times. There were incessant wars waged in ancient China, through the various imperial dynasties to modern times. Even during the Chinese Republic from 1911 to 1949, there were wars with Tibet, Mongolia, Japan, and Manchuria. There were also internal struggles, revolts, and civil wars, culminating in the war between the Kuomintang and the Chinese Communist Party, which came to power under Mao after a protracted revolutionary war.

India likewise suffered numerous wars of conquest and enslavement. The Hindus imposed a caste system on millions of Dravidians, who were mistreated as "untouchable" and denied equal rights. The Moguls conquered northern India in the sixteenth century. They occupied Delhi and large areas of India for four centuries. Their control ended in the mid-nineteenth century when the British East India Company established its hegemony and incorporated India within the British colonial empire. India achieved independence only after the Second World War. It was said during its heyday that the Sun never set on the British Empire—included under its rubric were colonies on all five continents. The British Empire collapsed after the Second World War, when it was unable to exert its global dominance any longer.

THE DECLINE OF THE FRENCH EMPIRE

It is instructive to focus on what has occurred in France in the twentieth century, particularly after World War II, in order to illustrate the role of contingency and turbulence in human affairs. We are close enough to these dramatic events to see what they mirror about the human condition in history, though any slice of human history will reflect the same phenomena.

The French colonial empire began in the 1600s and lasted until the 1960s, when it lost most of its possessions. At its height, France had a formidable army and navy and a powerful economy. French ships explored the high seas and established colonies everywhere. This was true, of course, of other European nations, notably Great Britain, Spain, Portugal, Belgium, and the Netherlands. The French colonial empire was second only to that of Great Britain, extending all the way from Indochina (Vietnam, Laos, and Cambodia) in Asia, to parts of Canada (Quebec and Nova Scotia) and the United States (Louisiana and New Orleans). It included many islands in the Caribbean Sea (Martinique, Guadeloupe, Guiana, Saint Lucia, and Haiti); it held Madagascar in the Indian Ocean, and it acquired many colonies in North Africa (Algeria, Tunisia, Morocco). In the Middle East it controlled Syria and Lebanon (Beirut was considered "the Paris of the East"). It extended its rule to French Equatorial Africa (Gabon, the French Congo, Cameroon) and West Africa (Senegal, Point d'Ivoire, Mali, Benin, Togoland, Niger, Chad, Upper Volta, Mauritania, French Somaliland, and its Central African Republic). It also possessed islands in the Pacific, such as Tahiti and New Caledonia. What an extensive empire!

The question that I wish to raise is: *Why* did the French colonial empire collapse after World War II, and what does that tell us about historical change in human history? France had been in constant battle with Germany going back to the days of war with Napoleon in 1870 (which Germany won) and continuing with World War I, in which France was victorious in 1918.

The loss of France's colonies was due primarily to the country's defeat by Germany in World War II. After World War I, France and Britain were concerned about Hitler's rapid rearmament of Germany and the strong possibility of another war, which they attempted to avoid. Édouard Daladier of France and Neville Chamberlain of Britain conceded to Hitler's demand at Munich in 1938 for the partition of Czechoslovakia. This allowed Germany to occupy the Sudetenland, which had a large German population. Hitler next turned on Poland on September 1, 1939. Britain and France learned that their policy of appeasement had failed and that they must now resist this aggression. They declared war on Germany, but Poland was quickly conquered by a blitzkrieg of Panzer tanks and Stuka dive-bombers, which terrorized the civilian populations. The Nazi-Soviet pact signed on the eve of this attack in 1939 made it amply clear that little could be done to save Poland, and the Soviets invaded Poland from the east.

There was an interlude of several months, which was called a *sitzkrieg*, or "phony war." This came to an abrupt halt on May 10, 1940, when Hitler invaded the Netherlands, Belgium, and Luxembourg. The French and British dispatched their armies to assist the Low Countries, which quickly surrendered. France had counted on waging a defensive war, for it was fearful of repeating the costly trench warfare of World War I. The German population was larger than France, and its birthrate was higher. The French were not eager to lose another generation of young men. They had built the Maginot Line, which was hailed as an impregnable fortified series of pillboxes and bunkers. Unfortunately, the generals of many armies are all too often predisposed to wage the *last* war. The same thing was true of France. The Germans had likewise constructed a strong, fortified Siegfried Line on its border, but Hitler's strategy was to fight an offensive war, using tanks and aircraft to punch through their adversaries' defenses. The French thought that the Germans would attack on the Northern Front, hence they rushed to the defense of Belgium. They were shocked when Hitler launched his offensive through the Ardennes Forest. This sudden thrust cut off both the French and British armies; many troops surrendered, while others beat a hasty retreat to Dunkirk on the Atlantic coast.

The Germans again surprised the French command, led by General Maurice Gamelin, when they turned south and east in order to invade France itself. The battle for France began and ended quickly. The French general staff was by now demoralized, its defeatism overwhelming. The German Wehrmacht had outflanked the Maginot Line by penetrating behind it and making it redundant. General Gamelin was replaced with General Maxime Weygand by the prime minister of France, Paul Reynaud. By this time the British and French forces had retreated to the channel ports and defended an enclave at Dunkirk. Hitler never expected the British army to escape his clutches, but large numbers of men were boldly evacuated by boats to England under cover of the Royal Air Force (RAF).

The devastating defeat of France now seemed inevitable. Although the French and British forces had been evenly matched with the Germans in infantry and tank divisions, the German air force was superior to the Anglo-French air forces, and it inflicted heavy casualties. German military strategy had outfoxed both the French and British high commands. The Allies had misjudged the German plan of action. Choices are decisive in human affairs, and the wrong policies can have terrible consequences.

Meanwhile Italy, under Mussolini, declared war on France on June 10, 1940, and seized the French Riviera (Côte d'Azur). The situation seemed hopeless. The French wished to save Paris, its historic monuments and boulevards, and so no defense was mounted and Paris was declared an open city. The French government hastily retreated to Bordeaux on the Atlantic coast. Paris was occupied by the Germans on June 14. France surrendered on June 22; an armistice was signed in the Compiègne Forest in the same railroad carriage in which the armistice of 1918 was signed. Hitler gloated over his spectacular victory; he was shown on newsreels dancing a jig!

The German blitzkrieg was devastating to French pride, which was thoroughly humiliated. Prime Minister Reynaud, who refused to capitulate, was replaced by Marshal Philippe Pétain, the military hero of World War I, and France was partitioned. The Germans occupied the north and west coast areas of France on the Atlantic Ocean, and a new government was installed in Vichy to govern the rest. The Vichy government attempted to collaborate with Germany; after the war Vichy was thoroughly discredited.

General Charles de Gaulle, who had warned France before the war that the next war would be motorized (he was ignored), had commanded a tank unit during the war. On June 18, 1940, he delivered a radio address in which he refused to accept the authority of the newly installed Vichy regime. He declared his intention of leading the Free French Forces and continuing the battle. French Equatorial Africa and Guiana allied themselves with de Gaulle's Free French Forces, though the colonies in West Africa and Indochina remained loyal to Vichy. France had promised Winston Churchill, by now prime minister of Great Britain, that it would scuttle its fleet. It did not, and many ships in the French navy were bombarded and sunk by British battleships to keep them out of German hands. Charles de Gaulle played a commanding role in France after the war. He proved to be right in his tactical support of mobile forces; he summoned the French people to resist the Nazi invaders without compromise.

Meanwhile, Britain was preparing to defend itself against a possible German attack. The rest of the story is well known. America entered the war on Britain's side on December 7, 1941, after the Japanese attack on Pearl Harbor. The Allies invaded France in 1944. Germany was soundly defeated by being squeezed between the Allies on the Western Front and Russia on the east. There were tens of millions of casualties. France was liberated in 1945 by the Allies, and although demoralized by its defeat, it was grateful to the American and British armies, along with the rather weak Free French Forces under the command of General Charles de Gaulle.

CHARLES DE GAULLE

Charles de Gaulle was a tall man (he was over six feet tall) with a big nose and a commanding presence with the demeanor of *hauteur* (haughtiness). He dramatized the role of the hero in history and refused to concede any diminution of French pride, in spite of the earlier military defeat by the Germans. From his redoubt in Britain during the war he had reminded the French people that he and the Free French Forces would return someday to liberate France.

De Gaulle was a rare figure, because he seemed to reflect high character, a man above the scandals of previous regimes. Although his politics were conservative and his domestic policies rightist, he was concerned primarily in restoring the status of France and its role on the international scene. Unfortunately, from his viewpoint, he was unable to save the declining colonial empire, in spite of valiant efforts on his part, but he was above all a realist in the competition for power on the international scene.

France was exhausted by the Second World War and could no longer hold onto its far-flung empire. It lacked the financial and human resources to maintain its hegemony. It did not have the economic or military power to do so. This became evident during the Cold War when the United States and Russia became the predominant superpowers, though the formerly powerful Soviet Union was no match for American power, had to withdraw from its military occupancy of Eastern Europe in 1989, and was unable to keep pace with the technological military power of the United States. Although de Gaulle attempted to maintain the glory of *La France* and some independence of its forces, it was clear to subsequent French governments that it could not restore the French empire to its past position of eminence. Roosevelt and Churchill had allowed de Gaulle to have a place at the peace table with Germany so as to assuage French pride, and indeed the Free French army under de Gaulle marched up the Champs-Élysées at the liberation of Paris, but this was primarily for show. It did not reflect the reduced role that France exerted in the invasion of Normandy, for the French army had surrendered to Hitler, and its fleet had been sunk.

What happened to the French colonies after the war is a story of further French military defeats and retreat. The French government tried to hold onto its colonies. It began with two wars: the War of Indochina and the War of Independence of Algeria—and it ended in ignominious failure.

THE INDOCHINA WAR

Vietnam has been throughout its history either an independent kingdom or a province under Chinese control. In the mid-nineteenth century, it was taken over as a colony of France. During the Second World War, the Japanese invaded Vietnam (in 1940), which was by then under the Vichy regime. The Japanese occupation lasted until March 1945, when the French returned to Indochina hoping to reestablish its colonial control; it was said that Indochina was essential if France was to revive its economy after World War II. As the Japanese left, the Chinese government, under the Kuomintang, was authorized by the Potsdam Agreement (of Roosevelt, Attlee, and Stalin) to occupy northern Vietnam in order to repatriate the Japanese army.

Meanwhile, the Viet Minh declared their independence from France and established their capital at Hanoi, under the leadership of Ho Chi Minh, a communist who had studied in France. The French declared their intention to continue their control and installed a puppet ruler, Emperor Bao Dai, in Saigon in the south. France assembled an army to consolidate its hegemony. This army included French troops supplemented by soldiers from its colonies, such as Algeria, Morocco, and Tunisia, as well as the Foreign Legion.

It became clear right off that the Vietnamese communists in their war of independence would resist French efforts to reoccupy the country. In November 1946 French naval vessels bombarded the port of Haiphong and killed several thousand civilians. In 1949 France recognized Vietnam as an independent state, though still within the French union. Meanwhile, the Chinese communists under Mao were victorious in their civil war. They began to provide arms to the Vietnamese. A fierce battle under Vietnamese general Vo Nguyen Giap began for control of French bases in the north. The war had been a rural insurrection at first. With China now under communist control, the Viet Minh were heavily armed with artillery and other modern weapons. In the northern part of Vietnam, bordering on neighboring Laos and Cambodia, the French army intensified its effort to defend its positions but suffered heavy casualties.

The Americans had continued to supply the French forces with financial aid. The Viet Minh forces invaded neighboring Laos; to counter this, French general Henri Navarre decided to make a concerted effort at Dien Bien Phu in northern Vietnam, near the Laos border. Some 20,000 French soldiers held the city but were defeated by 100,000 Vietnamese; there were heavy casualties on both sides. Some 2,200 French died in the battle; many were wounded and the rest captured (the Vietnamese lost 8,000 dead and 13,000 wounded). These casualties were very heavy to bear. (Incidentally, my wife's brother-in-law, Fred Kugel, was sent with the French army to Vietnam. He was a French citizen from Alsace-Lorraine. He recounted to me that he had lost his leg at Dien Bien Phu, the last major battle in Indochina. His leg was replaced with an artificial limb, and he walked with a limp. He recalled the desperate situation the French forces suffered.)

The French petitioned the United States for increased aid. When President Dwight Eisenhower refused to provide further support, the French decided to withdraw from "the dirty war." French premier Pierre Mendès-France, who had opposed the war since 1950, decided to enter into peace negotiations with the Vietnamese. The war had been costly to France, which was still recuperating from the Second World War. A peace conference was convened in Geneva on July 21, 1954, to arrange terms. Vietnam was divided at the seventeenth parallel. North Vietnam was controlled by the communists from Hanoi, and the Republic of Vietnam in the south was established in Saigon. This effectively ended French domination of Indochina.

The leader of North Vietnam was Ho Chi Minh—although a communist, he was also a nationalist who wished to maintain Vietnamese independence. The United States was fearful that the communists would control all of South Asia, so it had provided financial aid to the French to prevent that. After the French withdrew, both John F. Kennedy and Lyndon Johnson decided to recommit American troops. The second Indochina War (comprising Vietnam, Laos, and Cambodia) lasted from 1959 to 1975. It included the Viet Cong from South Vietnam and the North, as well as the Vietnamese Army and the Army of the Republic of Viet Nam (South Vietnam). What the Vietnam wars demonstrated was that it was difficult to prosecute a modern military action against guerrilla warriors fought in the jungles of a country. This was a lesson that both the United States and France were to learn. The United States, like France, eventually decided to withdraw from Vietnam.

One may ask: Why did the French lose Indochina? Two key events had occurred to weaken France's control. First, the United States' attention and aid were diverted when it entered a war with Korea (in 1950, under Harry Truman) to prevent Chinese expansion; and second, the French had to contend with a new war of liberation that broke out in Algeria.

Suffice it to say, successive French governments of the Fourth Republic tried to hold Indochina, but they could not because of the great financial drain, the heavy casualties suffered by French forces, and the intense opposition by both the French Communist Party and the Left in France. Indeed, in the early 1950s, there was increasing opposition to the war throughout France. It was politically difficult to continue military action.

Let us move on to the Algerian War of Independence, whose loss proved to be even more devastating to the French.

THE FRENCH-ALGERIAN WAR

The unraveling of the French Empire after the Second World War forced the French public to confront a series of traumatic events. The loss of Algeria was no doubt the most dramatic. Much larger in size than Tunisia or Morocco in North Africa, its proximity to France made the rupture between Algeria and France all the more painful. This was because a minority of the population in Algeria was of European descent, many of whom had lived there for a century or more. A significant portion of them were Sephardic Jews, by then French citizens. They either came after the Spanish Inquisition or could trace their origins as far back as Roman times. These French Algerians held most of the land, businesses, and power in Algeria.

The conquest of Algeria began in 1830 when Marshal Thomas Bugeaud, commander of the French army, invaded the country, no doubt an unjust effort to seize an area by force. Algeria was annexed to France in 1834. By the Second World War Algeria was considered part of France, and *l'Algerie Français* was loudly proclaimed. Indeed, Algeria was divided into three departments: Algiers, Oran, and Constantine. This was similar to other departments in France, though the Muslim population never possessed the same rights as their Christian or Jewish counterparts. Many French Algerians were truly accomplished, especially after they later emigrated to France—such as the well-known novelist Albert Camus, the playwright and novelist Emmanuel Roblès, philosophers Louis Althusser and Jacques Derrida, and the world-famous fashion designer Yves Saint Laurent.

On November 1, 1954, the National Liberation Front (FLN), under Ahmed Ben Bella, launched a guerrilla war in Algeria to gain independence. They demanded diplomatic recognition at the United Nations as an independent state. This came right after the defeat of France in the Indo-Chinese War. Although fighting began in the sparse countryside, it spread to the cities and became especially bloody in Algiers (1956–57), where there were constant violent bombings and urban attacks. The French forces responded to the insurrection by committing four hundred thousand to five hundred thousand troops and going all out to crush the FLN. Having been defeated in Indochina (the peace treaty had just been concluded in Geneva), the French nationalists thought that they had to make a strong stand in North Africa. Having surrendered to Germany in World War II, and having lost Indochina, France's national pride was at stake. The ferocity of the conflict increased as French forces desperately resorted to terror and torture to defeat the FLN, including the first use of helicopters and napalm on civilians. This was in response to FLN guerilla warfare in which no one felt safe. The Algerian rebels were no angels, and they resorted to constant bombings, ritual mutilations, and kidnappings to terrorize the French Algerian population.

(On a personal note, I visited France in 1958 at a time when a great controversy was raging in *Le Monde*, *Le Figaro*, and on the streets of Paris. Many held that Algeria was French and needed to be kept no matter what the cost. Others thought that the war was too brutal and that by using terror and torture, the French army betrayed the Rights of Man. To add to the intense emotion, my wife's grandparents, Ferdinand and Louise Vial, had lived in Oran for thirty-five years. They owned a small hotel (*pension maison*) and restaurant. They were forced out in 1962 and resettled in Mouans-Sartoux, the original family hometown, going back centuries. Similarly, her aunt, Therez Cotrez, who lived in Oran, was also compelled to leave. They lost everything but their valises. My wife had lived with her grandparents in Oran for two years and has many stories to relate. I was overwhelmed by the terrible toll that the war was exacting, not only on Algeria but on the French nation, which was deeply split on the morality of the war. In subsequent years I met many *pied noirs* (the former French residents in Algeria) in Southern France who never got over their defeat and expulsion from Algeria.)

The loss of Algeria on top of the loss in Indochina had a devastating effect on the French psyche: the *gloire* of France was now overshadowed by the power of America, which had liberated France but had also emerged as the superpower, far more resourceful than France, which was watching as its empire disappeared.

But to return to the story . . .

The French political system was thrown into disarray and was unable to cope with the conflict; government after government fell. Meanwhile, there was widespread criticism of colonialization, not only within France itself but by worldwide public opinion. Many saw a contradiction between the French commitment to the Rights of Man that had inspired the French Revolution and the atrocities perpetrated by the French army on civilian populations in response to terrorism. There was a significant minority of Christians in Algeria (i.e., the European settlers), but the majority of the population was Islamic. The French Communist Party, though opposed to "capitalist imperialism," sought to reach an accommodation, as the Cold War was now in full swing and the party was more interested in that conflict. However, a number of French intellectuals, especially Jean Paul Sartre and Simone de Beauvoir, were scathing in their attacks on France's role in the war. There were also disquieting non-Marxist objections coming from Roman Catholics and other writers. A number of intellectuals issued *Manifeste des 121* in support of the dissenters to the war in Algeria.

A coalition of *pied noirs* and army officers set up a committee of public safety to challenge the French nation. They called upon General de Gaulle to return to power and save France. He returned to set up a Fifth Republic and to write a new constitution, which allowed for a president to be elected for a seven-year term, who could serve in spite of parliamentary instability. This was supported by a huge majority of the electorate, who continued to proclaim, "*Vive l'Algérie Français!*"

I remember visiting Paris during a military parade in which the commanding figure of de Gaulle participated. At that time he meant to restore the grandeur of France with a strong show of force. At first he pursued the war with vigor, yet at the same time he was open to the possibility of a peaceful agreement, for he understood how determined the Muslims in Algeria were in becoming independent, and he felt that he could no longer oppose that successfully. De Gaulle apparently was deeply concerned about the demographical disparity between the rapidly growing native Algerian Muslims and the slow birthrate of the French settlers of European stock. If Algerians were to remain French citizens, in time they would outstrip the *pied noirs* in numbers and voting power.

On September 16, 1959, de Gaulle suddenly reversed his position and proclaimed the right to Algerian self-determination, hoping the Algerians might join the newly established "French Community." It was clear that by now the overwhelming majority of Islamic Algerians preferred total independence from France.

This infuriated the *pied noirs* in Algeria, who allied with a coterie of right-wing army officers, protested that this was an act of betrayal, and attempted a *putsch* in April 1961. They were part of a secret army organization, Organisation

Armée Secréte (OAS), which was committed to defeating the Algerian separatists. Their attempt to overthrow the government failed. De Gaulle decided to abandon the recalcitrant *pied noirs* entirely, and his government entered into peace talks with the FLN. French voters approved Algerian independence in a referendum in June 1962. This was followed in July 1962 by a referendum on independence, which was overwhelmingly approved by the Algerians.

In the following year, some 1.4 million refugees began an exodus from Algeria. Included in this were several thousand Muslims who were pro-French. A considerable number of Algerian soldiers had fought alongside the French during the war for independence. These were known as *Harkis*. Many had fought with France in Indochina and other campaigns. Approximately 220,000 Harkis were disarmed when the French left. Although the FLN had agreed not to harm them, an estimated 150,000 were considered traitors and were massacred. About 40,000 managed to escape to France. Only forty years later did Jacques Chirac (former mayor of Paris and president of France), who had fought in Algeria, apologize to the Harkis for not being welcomed in France. The Harkis now number almost half a million. Indeed, France's Islamic minority is by far the largest in Europe, approaching five to seven million.

So ended a bitter chapter in French history, as its colonial empire was defeated by costly battles. Other parts of the empire were also decolonialized. Black Africa, a huge landmass, was under French colonial control. Here the French withdrew gradually without a protracted civil war (with the exception of Madagascar). One reason was that there were insufficient guerrilla forces capable enough to embark upon a war of insurrection against the French.

THE CONFLICTS BETWEEN NATIONS

Perhaps it is time to reflect on the role of contingency in France, Indochina, and Algeria as illustrative of the dramatic conflicts between nations. What happened was a result of developments in the so-called superstructure of the political and ideological factors in both the colonies and France. This was influenced by the broader Cold War between the Western democratic countries and the Soviet and Chinese Marxist bloc.

The simmering political conflict within France was between socialist and communist parties—often at loggerheads with each other—and the old rightist parties in France. The emergence of General de Gaulle, who had headed up the Free French Forces against the Vichy, and his effort to restore France to its former position of eminence in the world, was also a key factor. This French hero, like Napoleon a century and a half earlier, had played a decisive role at key turning points of this controversy.

After the war France tried to maintain some kind of economic, cultural, and political relationship with the Francophone sectors of Africa. It provided favorable economic terms to stimulate and maintain trade with these countries. At first the currency of the Francophone African countries had a favorable rate of exchange with the French franc (two to one) in order to stimulate French investments in these countries, but it could not provide massive investments, and it allowed each of these countries to become independent. France continues today to station a regiment of crack French troops in Gabon, but it does so because it is the only French-speaking former colony that has struck oil and in which French oil companies have heavily invested (Elf Aquitaine). Once the euro replaced the franc, the currency advantage no longer existed. Meanwhile, other countries have replaced France with investments, notably the Chinese government, whose presence is increasingly felt because of its need to guarantee the natural resources of these countries in the future.

Perhaps the most important relationship of the former colonies to France is the linguistic affinity, because French is the official language of the governments and of the elites, many of whom were educated in France. The Francophone areas of former French colonies include the province of Quebec, which is now independent of France and an integral part of Canada. Similarly for scattered islands, such as Martinique and Guadeloupe in the Caribbean, Madagascar in the Indian Ocean, and Tahiti in the Pacific Ocean; also the former French colonies of Lebanon, Syria, Algeria, Tunisia, and Morocco. The French are horrified by the decline in influence of the French language, which had been and still is the language of diplomacy at the United Nations and other international bodies. But English has now become the international language of choice, a requirement in any nation that wishes to maintain its economic and political rule. In former days the elites of the world would educate their children in the French language, and there were *lycèes* in most of the major capitals, where French was taught. Now *Franglais* has overtaken the French language, as English technical and economic terms and *argot* (slang) have been incorporated into the French language.

Unfortunately, the popular great French cruise ships that crossed the Atlantic Ocean before the war, such as *Le Normandie* (which was sunk by sabotage in the Hudson River during the war), *L'ile de France*, and *Le France* (which was sold and renamed *The Norway*), no longer exist. Formerly these vessels were constructed in the bustling port city of Le Havre, which also ceased to be a shipbuilding port. France does maintain a small cruise line, the French *Club Med*, which comprises fairly modest French yachts and sailboats. French rule as a sea power had been an integral part of its international place among nations, but this is no longer the case.

The decline of French military power is another reason why its international influence has waned. France maintains a crack air force and produces *Le Mirage* fighters. It has a modest army, but much of this is integrated into the North Atlantic Treaty Organization (NATO) as part of a European force. Hence, France has largely withdrawn from the former bases it maintained worldwide.

French economic power makes it the *sixth* largest economy in the world in terms of its gross national product. It is still an important financial power in the world because it maintains a stock exchange (*La Bourse*); however, it cannot match the economic resources of the United States, Japan, Germany, the United Kingdom, or the rising power of China and several emerging economies (such as Brazil), which are likely to overtake it. France is best known for its export of fine agricultural products, from cheese to wines and liquors, and its reputation for women's fashions and luxury goods.

France's role in the Industrial Revolution of the eighteenth and nineteenth centuries was significant, and in the twentieth century its mines and factories produced fine automobiles, aircraft, clothing, and other important products. France has followed other countries in losing its manufacturing base, because goods can be produced much more cheaply abroad, where the wage scale is lower, so goods from China and the third world are increasingly replacing French products. One interesting point is that the lion's share of electrical energy is produced by nuclear power plants by the French-owned *Fondation d'Electricity*.

Having no natural oil or gas reserves, France is forced to import all of its carbon fuels. It decided virtually before anyone else to develop nuclear energy, and it has built clean and efficient plants, which give it a decided advantage over other European powers.

After World War II France was the leading proponent of creating the new political entity of Europe. One reason for that was to end the military rivalry with Germany by integrating it into the European Common Market and the European Parliament. Actually, a key advocate for this was Jules Moche, who worked valiantly to make Europe a reality. Giscard D'Estaing, former president of France, drafted the European constitution, which traces France back to its humanistic origins in the Hellenic world as a secular culture, barely mentioning its religious and Christian past. There is still a long way to go to complete the ideal of creating an integrated Europe—and there have been and no doubt will continue to be obstacles placed in its way. But France is increasingly part of Europe, and there are joint projects that are manufactured, such as the Autobus, which competes with Boeing aircraft in the sale of airplanes. Similarly for CERN in Switzerland, which is the major superconductor in the world engaged in research in subatomic physics.

Of considerable significance is the fact that the gross national product of Europe (all of the countries combined in the European Common Market) is virtually equivalent to that of the United States. So France and other European economies have more clout as part of the European Common Market.

France maintains a kind of supremacy in two areas. The first is cultural—Paris for centuries was considered to be (at least by the French themselves) the center of intellectual life in philosophy, literature, and the arts, and at one time it played a significant role in science. It was considered avant-garde in many fields. Politically, many of the revolutions of the world, it was said, were spawned in French *cafés* and *brasseries*. For decades France was the magnet for left-wing movements, at least until 1968, according to French philosopher and journalist Raymond Aron. It had a powerful Communist Party (no longer) and strong labor unions. Ever since de Gaulle established the Fifth Republic under a new constitution, France has become fairly stable politically—formerly governments of the Third and Fourth Republics would fall incessantly. Now there is a president elected for a seven-year term, and the conservatives vie with the Socialist

Party for power. Nonetheless, France still has the reputation of being on the cutting edge of *la nouvelle vague* (the new wave). Surely, the sudden appearance of existentialism after the war—Jean-Paul Sartre, Simone de Beauvoir, and Albert Camus—gave that impression, as did the impact of postmodernism, influenced by Jacques Derrida and Jacques Lacan, and structuralism, influenced by Claude Leví-Strauss. Increasingly one hears in France many complaints that Paris is no longer the center of the universe and that New York City, London, Tokyo, and Beijing have replaced the French appeal. One problem that France faces is the fact that it is difficult for young French students to find attractive jobs or careers in France, and many have moved to London or New York City.

French cultural influence continues in the visual and performing arts. Modern art surely began in Paris (Picasso, Braque, Matisse, and Manet became known worldwide), though increasingly people talk about it moving to New York City or London. Surely France maintains its supremacy in cuisine (*le Guide Michelin* indicates three-star restaurants) and haute couture (high fashion for women), so its impact on culture is enduring, though on a lesser scale. The quality of French taste still sets an example to the world.

Another dramatic change in France is the nigh total collapse of the influence of religion, especially Roman Catholicism, in French life. Indeed, France has become highly secularized, and the vast majority of French men and women are largely indifferent to religion or will participate only at weddings or funerals. The main emphasis seems to be on the *joie de vivre*. Of course, ever since the Avignon Papacy, when there was a second competing pope, the French have been a thorn in the side of the Vatican. But France, as a result of the rationalism of René Descartes and the *philosophes* of the Enlightenment (Voltaire, Diderot, d'Holbach, and others), has inspired many modern thinkers. Moreover, French cynicism makes it difficult to impose rigid rules, particularly concerning marriage and sexuality. The French are notoriously astute in perfecting affairs with mistresses and lovers. One has only to read Gustave Flaubert's classic novel *Madame Bovary* to see how it was done. And so for the Frenchman and Frenchwoman personal liberty is preferred to repression. The French Revolution with its heralding of "the Rights of Man" (*and* Woman) still inspires countless generations of modernists.

A significant change in French life is the growing segment of the population that is Muslim, the largest such minority of any European country. Many critics are fearful that Muslims will outbreed everyone else and change France's ethnic makeup. Optimists say that this is unduly pessimistic and point to the fact that young Muslim men and women are increasingly accepting the ideals of the French Revolution: "Liberté, Egalité, Fraternité," and they hope that in time they will fully accept the lure of French culture, become truly French, and intermarry outside of or indifferently to religion.

No doubt the largest industry in France is tourism; people from all over the world visit France as the number-one tourist attraction. They find the French cities, such as Paris, beautiful; the French countryside charming; the French Alps and seashores thrilling; and French perfume, cuisine, and women especially attractive! The French Empire of old has been transformed from a military, economic, and political power into a magnet because of its cultural contributions to world civilization.

In summary, France as a country has had a tumultuous past. It has experienced wars and revolutions, immigration and interchange. It has managed to survive and remain independent. There is a French saying: "*Le plus ça change, le plus c'est la même chose*," to which I respond, the more that France remains the same, the more it changes, and this is due to a multiplicity of causes.

The term *civilization* (or *empire*) is a broad concept, encompassing a rich tapestry of human institutions under it. But I think that the effort to explain virtually any and all human institutions—whether civilizations, empires, nation-states, city-states, villages, universities, churches, voluntary associations, scientific or cultural organizations—demonstrates that they all display similar dynamics. Each of these social institutions is composed of human beings, and its historic dimension likewise involved real human beings of the past. The "corporate" body that has emerged is composed of the habits, traditions, rules and laws, beliefs, and aspirations that have developed, and these are the adhesive that keeps the organization together and prevents it from splintering, which it may do at certain periods in its history. Nevertheless, the whole is more than the sum of its parts; it does not exist separate from the human beings or their conduct, beliefs, convictions, and values that justify the institution to which they are committed. And it continues its identity for as long as it is a viable, functioning institution because of the habits of conduct of the human beings who make it up. If the institution is appreciated by those who take part in it, its founders will be heralded and acclaimed, its founding documents cherished, and people will abide by the standards and values they set.

POSTSCRIPT ON THE USA

We can raise similar questions about the United States of America. Like France, it is a country with a history that takes on "mythic" proportions. People who are born in the United States or who become naturalized citizens revere its historical past and are loyal to the idea of "America the Beautiful." They may "pledge allegiance to the flag of the United States of America and to the Republic for which it stands." And officials who are elected to office take an oath or affirmation, reiterating their commitment. "I do solemnly swear (or affirm) that I will faithfully execute the office of President of the United States, and will to the best of my ability preserve, protect, and defend the Constitution of the United States."[6]

The people of the United States generally consider the Founding Fathers—Washington, Jefferson, Madison—to be noble men committed to the Republic. And they have added to the list of revered presidents Abraham Lincoln, Franklin Delano Roosevelt, and others who strived so valiantly to protect the Union and extend the blessings of liberty and equality.

The United States as a nation exists in the minds and hearts of countless Americans who cherish its goals of "life, liberty, and the pursuit of happiness," and it is a new "reality" in the nature of things, implicit in the determination of the people who believe in it and will conduct themselves in terms of its laws and ideals. The United States thus has some ontological status in the nature of things, much the same as Greece or Rome did, and as Spain or Mexico, Syria or Sri Lanka, China or Japan do today.

All such sociocultural institutions are emergents of the collective will and commitments of the people—past, present, and future—who have made or will make them up. Here turbulence and conflict, law and order, contingency and harmony apply in some meaningful sense. Moreover, to explain how and why such institutions function the way they do, one needs to examine a multiplicity of causes—political, economic, religious, military, sociological, psychological, and cultural. One need not invoke the "hand of God" or "divine destiny," nor reduce the country to the physical-chemical effusions in the brains of its citizenry to understand it or to cope with any problem that may arise, with the policies enacted, and with the officials elected to carry them out. What will happen to the United States depends in part on the decisions made about policies and laws by the electorate and the political leaders. If we briefly review the history of the United States, we see that the constant theme has been one of change and flux.

The North American continent was first settled by immigrants coming over the Bering Strait from China and Mongolia some ten thousand to fifteen thousand years ago. European colonists began arriving only yesteryear, in the fifteenth century. They defeated the native Indians and expelled them from their lands in their effort to conquer the vast continent. There were only three million inhabitants of the American colonies when they issued a Declaration of Independence in 1776, separating themselves from Great Britain. They divorced themselves not only from Britain but also from the colonial powers of France and Spain, and they established an independent republic. After the American Revolution they wrote a Constitution, which was enacted in 1789, and a Bill of Rights. The Africans who had been seized and enslaved, along with their children, were eventually emancipated in 1863 by Abraham Lincoln as a consequence of a bloody civil war. In the nineteenth century extensive industrialization occurred; the United States became the preeminent nation on Earth, trading worldwide. It opened its doors, beckoning people from all over the world to come to its shores seeking a better life, and by the beginning of the twentieth century it had one hundred million inhabitants. The United States became the most powerful nation in the world and embarked on wars and alliances everywhere—the Spanish-American War, the First and Second World Wars, the Vietnam and Korean Wars, the Cold War with communism, and the wars with Islam in Iraq and Afghanistan. Its scientific and technological discoveries; secular universities; free religious churches, temples, and synagogues; and democratic form of government gave it impressive vigor and attraction. As the country entered the twenty-first century, with three hundred million strong, it was the dominant global power, which aimed to be virtuous but was considered by its critics to be overweening in its ambition.

The leader of the free-market economy on the planet was often in trouble, an open pluralistic society always in need of reform to protect the disadvantaged from the wealthy. It was among the first to explore outer space, the Moon, Mars, and the various planets within the solar system and beyond. It was a nation focused on improving the lot of all its citizens, while at the same time it defended repressive regimes throughout the world. For every generalization made about the United States, there are counter-generalizations.

Given the information revolution, the United States played a pivotal role in global communications, and due to a free press controlled by plutocrats, yet more or less free, it was constantly forced to make choices, by itself or in cooperation with other nations, as it confronted new challenges and problems. Whether it succeeds depends upon the policies that it adopts and the decisions that it makes, and whether they are made with wisdom.

There were economic crises faced by every generation—the panics, recessions, or depressions of 1837, 1857, 1873, 1893, 1907, 1929, 1938, 1981, 2008—and the efforts to deal with them. And there have always been political and military conflicts that demand appropriate responses. Both the United States and France have faced awesome challenges. One can never stop the roller coaster in mid-ride in an attempt to get off. As a matter of fact, what will be in the future is not preordained but depends on the choices of those charged with deciding, and this depends on the policies adopted by the economic and political leaders, and in democratic societies by *le grand public*. There are always debates going on; and there are those who must be responsible for decision making either within their own personal lives or on behalf of the political-economic elite. A nation's leaders cannot escape from making choices, and these, along with other factors—internal and external—determine its future.

Accordingly, the moral decisions of the leaders of France, the United States, or any country are causative. Given turbulent forces, conflicts, and disagreements, questions are raised: Can we actually make choices? Do we have the power to determine the future? Should these be left to blind forces, or can we use reason to make wise choices? I submit that a nation has no other choice but to choose. Even not to choose is an act determining the future course of events. If we seek to escape from freedom (as Erich Fromm observed for an earlier generation) and affirm that *only* God will save us (i.e., the priests or mullahs), or that only the codes inscribed in ancient documents (the Hebrew Bible, the New Testament, the Koran, or the Hindu beliefs and practices) can help us—then we have abandoned any effort to control our destiny and the future of our country or civilization.

If the universe shows no purpose or order, no meaning or plan of salvation, then it is an open, pluralistic scene, and we need to make the best-informed judgments that we can. What will be depends on contingent facts *and* rules, principles, rights, and duties that have evolved in a culture and its constitutions and rules of governance—the best collective wisdom of the time.

I submit that in the course of human affairs a series of principles, norms, and guidelines have emerged, and these may be characterized as *planetary ethics*, incorporating the wisest and most sensible norms of cultural behavior that humankind has evolved—the fruits of the long road that human civilizations have taken to ameliorate the human condition. These grow out of the collective human experience and the significant set of ethical values and principles that have emerged—and it is to these that I wish to turn. Whatever is said about the United States and France can be said of any country on the planet. Their futures depend on many factors, including the economic and political choices of their leaders and the public from which they derive their authority—and whether these decisions are based upon prudential wisdom. But it is important today to place these principles and problems in a broader planetary context.

NO DEITY WILL SAVE US. WE MUST SAVE OURSELVES!

All of the above recommendations make eminent sense from the short-range standpoint of democratic market-oriented affluent societies, including emerging economies that aspire to achieve similar aims. At present this is the dominant goal of most nations in the planetary community.

This says very little about the midrange problems that our planetary civilization will no doubt face—namely the severe environmental and population problems facing the globe. Humankind has limited resources to use or squander; these are finite and many are not renewable. Coal, gas, oil, and other natural resources someday will be exhausted. Industrial-technological societies have used these with abandon, squeezing the Earth dry to satisfy their greed for economic growth, no matter what the long-range consequences.

Global warming and other environmental strains on the planet will no doubt challenge human societies in the future to adopt radical measures, if they are to avoid disaster. One may only hope that human intelligence will contrive bold policies to overcome environmental depletion using renewable resources. It is conceivable that humankind will muddle through these crises, especially if new forms of energy are discovered and if and only if we can transcend the limited horizons of purely self-interested nation-states by developing transnational institutions to restrain waste; that is, by creating a democratic world federal system.

This says very little about the *long-range prospects* of the human species on the planet Earth. These are largely unpredictable. What if sunspots flare up on the surface of the Sun? What effects will this have upon life, particularly human life, on the planet? What about other possible cosmic reverberations?

Former periods of warming or global ice ages may have had astronomical causes that were difficult to anticipate or deal with; similarly for unseen bacterial or viral infections, such as the plagues that decimated populations in the past. The long-range ultimate prospect of *Homo sapiens* on our planet and solar system is unknown. Perhaps these can only be understood if and when such contingencies occur. At some point in the future, humankind will have to cope with unforeseen possible threats. Humans have risen to the challenges in the past; hopefully they will do so again in the future. This will depend on humankind's resolve to marshal the best resources—intellectual and moral—to succeed. It will also depend on whether humankind will be able to summon the courage without retreating into a cocoon of unreality by postulating unseen divine forces to save humanity. To which we need to invoke the secular imperative: "No deity will save us; we must save ourselves!"[7]

This resolve can only happen, I submit, if we are able to transcend the moral creeds of the past, drawn from ancient religions that offer solace to grieving hearts. If the problems of humankind are to be resolved, then we need to use the methods of scientific inquiry, and we need also develop a new humanist ethics that is relevant to the human condition, not mired in supernatural fantasy or out of cognitive touch with the world of nature.

It is clear from our examination of the biosphere that there is an ongoing battle for survival, and that conflict and dissonance are ever present in a random universe. Recognizing the violent biogenetic past of the human species, we are now obligated to draw up new norms of morality. With this in mind, there are three questions that we need to address:

First, do humans have the power of choice?

Second, which ethical principles will help us to ameliorate the human condition?

Third, what can we do to apply these to the planetary civilization that is emerging?

ACT EIGHT
MORAL CHOICES IN A RANDOM UNIVERSE

INSPIRATION AND INTELLIGENCE

There are several human traits that are clearly favorable for survival of individuals and social groups, notably the development of *coping intelligence*, the *courage to persist* in spite of obstacles, the *willingness to work hard* to achieve set goals, some measure of *cooperative* behavior in social groups or between them, and the *defense of the young, their nurturing and education.*

Another key factor is the *plasticity* of humans and their ability to modify their behavior in different sociocultural environments. If one compares the great multiplicities of cultures and the ability of individuals and groups to adjust to them, we must conclude that *adaptation* itself has become an ingrained tendency that can redirect genetic tendencies appropriate to the new contexts in which people find themselves. Living in an ice age's frigid environment is different from living in a torrid environment, most particularly in the kinds of clothing that people wear or the shelters in which they live—heavy skins or wools for warmth for living in cold climates, or skimpy clothing or nudity in hot climates.

There are several *I*s at work here:

- *Instinct*—the tendency to behave in a certain way, as genetically endowed
- *Inspiration*—the ability to come up with new creative ideas to cope with problems
- *Intelligence*—the powerful coping mechanism that humans use for their survival

Inspiration and intelligence combine to modify behavior in response to the threats or opportunities in the natural or sociocultural environment. It is the combination of inspiration and intelligence that enables humans to overcome obstacles and redirect instincts. Inspiration is related to two other *I*s:

- *Invention*—of new tools, instruments, and objects for use
- *Innovation*—of new departures in behavior and of new means introduced to modify and attain their ends

Both of these are related to human intelligence. The entire drama of human experience—from fifty thousand or sixty thousand years ago to the present—is the rapid acceleration of cultural changes that are transmitted to successive generations, particularly in the past 11,700 years. These are repeated by oral traditions, preserved by customs, and eventually written down to become an integral part of the cultural heritage of a society. In this manner the human cultural heritage is added to *Homo sapiens*' biogenetic endowment. *Coevolution* thus occurs. The biological framework of a species is transmitted by *genes*. The sociocultural traditions are conveyed by *memes*: the patterns of belief and practice adopted in a society and transmitted to future generations. Culture becomes an integral part of a civilization and provides some stability in people's expectations. What are considered to be the proper modes of conduct are enforced by priests, warriors, teachers, and through social rulers.

Nonetheless, although these habits may endure for some time, they are not eternal, and eventually they may themselves change or decay. Although civilizations may endure for extended periods of time and even be sanctified as eternal, like all other things in the universe, they are buffeted by the sands of time and eventually submerge or disappear. Cultural patterns and social institutions are ultimately fragile, even though they serve as the repository of revered beliefs and sacred values for periods of time. Civilizations, like all things, are contingent; they may crumble because of clashes with other civilizations, intermarriage with foreign persons holding different conceptions of propriety, the extinction of old-time religions, or the continued modification of languages infiltrated by other linguistic stocks.

What are the key ingredients of sociocultural change? First is the clash or integration with other cultures; second, unforeseen changes within the environment; third, the role of moral choices in behavior and the capacity of *Homo sapiens* to redirect the future course of events.

ETHICAL CHOICES

Human beings always wonder whether events that have occurred could have been otherwise. My response to this question is in the affirmative, though within limits; thus a major stimulus to sociocultural change is human *creativity* that combines inspiration and intelligence, innovation and invention, and the capacity of humans to modify behavior in the light of an intelligent evaluation of existing conditions. The creative introduction of new departures in behavior plays a special role in human affairs. Human beings are capable of *inventing* new tools and adopting new modes of conduct. Thus it is *eupraxsophic pragmatic intelligence* that enables us to make choices and adopt new forms of behavior, whether conformist or radical.[1] Men and women are decision makers bound by their instinctive tendencies, which serve them to fulfill their needs. Their capacity to make choices provides ways of adapting to challenges in the environment.

The study of history reveals the great diversity of languages in human cultures. There are an estimated eight thousand known languages. The evolution of language is like the evolution of all kinds of things encountered in the universe, separately developed by different societies—understood only by those born into a society and brought up under special sociocultural conditions. One can speak French, Spanish, or Italian—all Romance languages with a common Latin source—yet deviating depends on the geographical location. People can learn to speak German, Yiddish, Polish, Russian, Chinese, Sanskrit, Bantu, or any other polyglot creation. Similarly for the flexibility of cultures, their unique moral, economic, political, and sociological patterns develop historically in their own ways in splendid isolation, yet often they encounter other cultures, whether by peaceful commerce or conflict, and are modified in the process.

What is my point? There is great malleability and diversity in ways of living and transacting with others and a gradual or rapid change of cultures. For example, the French language developed as the language of the educated elites and of international diplomacy. Today it is becoming *Franglish* (which the French deplore), because it has imported English phrases and technical terms as part of a wider planetary civilization that is developing. It is English that has become the international language of trade and cultural exchange. This is no doubt due to the power of the Anglo American empire that has become dominant, and also of the new electronic forms of communication—mobile phones, the Internet, movies, and television—that are able to flood all sectors of the globe with the ideas and values, forms of taste and behavior, influenced by Hollywood, London, New York, Beijing, and New Delhi.

The most decisive illustration of the capacity of humans to adapt rapidly to new cultural conditions is immigration and emigration, whereby large numbers of persons are transplanted to entirely new lands and are forced to learn a new language and adopt new manners and customs. This of course goes back to the earliest days of human history, when invaders would capture foreigners and bring them back as sexual objects or slaves. This has always been true in tribal warfare, when invading hordes of male warriors would plunder the territory, abscond with the women and children, and force others into slavery. The Romans were adept at such practices; the defeated foreigners were forced to serve their masters and adopt their culture. The capturing of blacks in Africa and their enforced slavery vividly demonstrates the cruelty of such persecution and exploitation. African Americans were compelled by European colonists to serve their masters, which meant the separation of families and enforced servitude in a new culture.

This phenomenon of cultural implantation is all the more true today, as people freely leave their native lands in search of a better life in North and South America or Australia and New Zealand. Uprooted by choice, Italians and Germans, Russians and Chinese, Spaniards and Indians, Jews and Gypsies have migrated to new lands, abandoning the culture of their fatherlands and adapting to new sociocultural values. Impressively, foreigners once denigrated as "wops" and "kikes," "chinks" and "spicks" have become Americans or Canadians, Argentineans or Brazilians, and their children have been able to assimilate an entirely new culture in one generation.

Human beings are malleable, able to imbibe entirely new forms of behavior as they assimilate in distant cultures, learn their languages, and accept their mores—almost overnight. We observe time and again the readiness of people to adapt to a wild-frontier society as Europeans did on first coming to America or as foreigners do today—to an automobile culture in which they live in a suburban environment. My wife, children, and grandchildren are virtually Franco Americans, able to live in two countries, with a diverse set of values (our capacity to enjoy croissants and hamburgers, beer and wine, baseball and soccer, *La Vie en rose*, and ragtime jazz). Of course, this is because America and France have shared cultural values for centuries, but it also applies to radically different cultures—as shown by Nepalese, Bangladeshi, or Koreans who are able to adapt readily to their new American homeland and who often integrate and prosper.

So, what may we infer from this? There is a remarkable capacity of humans to make new choices, whatever the cultural context, and in spite of turbulent times, to abandon the values of the old-time religions and to move on to new forms of living, often in a relatively short period of time, as their children's children become rapidly assimilated into and absorbed by the new culture.

Thus, there is a constant flow, as civilizations come and go, but humankind has finally reached a new plateau—*the emergence of a planetary civilization in which all parts of the public community now take part*. There are fewer completely isolated breeding groups, separate cultural enclaves, or totally independent social systems. Clearly, there are pluralistic communities in the world, but they have an interdependent transactional existence. No community can live in totally isolated ignorance, indifferent to the needs and wants, beliefs and values of other humans: the planetary community, for the first time, shares common ground.[2]

And herein lies the key to the future of humankind. If nature and the biosphere as well as the physical universe and human affairs have heretofore been particularistic, separatist, and isolationist, this is no longer the case, because there are common interests and needs along with shared dreams and values that unite us. And now we have the technology to communicate instantaneously as part of a global village. The urgent imperative is for humanity to develop a new world order; it is forced upon everyone whether they like it or not.

This is *planetary ethics*. It is unlike anything that has existed before. The separate cultures recognize that though they are rich in historical diversity, they nonetheless share a common desire for a peaceful and prosperous future. Although the universe is turbulent, contingent, random, and chaotic, there is the recognition that we need to establish a new planetary moral universe, formerly "the brotherhood of mankind." Although nature is indifferent to the moral values of the human species—which is just one among other species on the planet—*we do care*, and there is a new *moral universe* that has been emerging. The human species is a product of coevolution. We possess a biogenetic endowment with instinctive tendencies for *need reduction and satisfaction*. But also our cultural evolution has been accelerating, particularly since the last ice age. There is accordingly a *moral universe*, in addition to the physical universe, and its location is in cultural civilization.

Indeed, the evolution of *ethical culture* provides the first opportunity for humankind to maximize the dimensions of freedom and the latitude to create a new ethic, drawing upon the ethical wisdom of the past for values and norms that are enduring (such as the common moral decencies), and also the capacity to fashion new ethical principles appropriate to the multisecular, transnational, and planetary world in which we live.[3] This includes the ethics of enlightened self-interest *and* a compassionate empathetic altruism. The difference between a *moral* and *ethical* culture is that the latter is *infused by reason*.

"WHAT IF . . ."

The occurrence of the improbable in the universe is especially evident in the biosphere, which is more like a gambling contest than a well-designed formal plan. In the biosphere species emerge and become extinct. Contingencies are evident in the physical universe; turbulence is all the more evident in the arena of human affairs where uncertainty is invariably present. The human *lebenswelt* is replete with indeterminacy and chance; good and bad luck can enhance or wreck our lives.

The question has been raised: Is the uncertainty principle only due to our ignorance of the underlying causes at work, or does it have a real foundation in nature? Is the natural world unfathomable because we cannot fathom all that is at work, or is nature itself *uncertain*? All the crows that we have encountered thus far are black, but perhaps a *pink* crow may appear, or a *real* mermaid. We need an open mind. Our knowledge is surely fallible and subject to self-correction.

A pink crow may appear some day as a mutant. The existence of lovely mermaids (half-woman/half-fish creatures in the realm of fantasy) is highly improbable, though we may encounter a persistent sea captain like Ahab from Herman Melville's *Moby Dick* who will not give up his quest for the white whale. Do we need to drain all the seven seas to find out finally if mermaids exist?

There are, of course, singular, bizarre, totally improbable events that do occur. Are they that way because nature is that way? True, in human affairs, almost anything can happen overnight, and we are dumbfounded when it does. The twin towers of the World Trade Center collapsed in 2001 in New York City. It was an improbable, unlikely event that burning fuel emitted from two 747 airplanes was hot enough to melt the supporting columns and that the buildings would collapse pancake-style. Some speculate that perhaps it was due to a conspiracy by the Pentagon and/or Massad (the Israeli intelligence service) to catapult the United States into a new war on terror. There is little concrete evidence for this conspiracy theory, though it is widely held.

Was the tsunami that suddenly struck the Indian Ocean in December 2005 and killed two hundred thousand innocent people a random event? Science tells us that it was caused by Earth's shifting plate tectonics, which caused an earthquake that resulted in a tidal wave. So there is a causal explanation that geological science provides to explain this event, which seemed so unpredictable beforehand, though other earthquakes have been explained by the shifting of tectonic plates. That all of those people trapped on the beaches and shores of the Indian Ocean would be drowned was totally surprising and unexpected. This tragic event was due to the impact of the tidal wave on human affairs, and here the contingent seems like a genuine explanation. We do encounter flukes and accidents, and luck and chance do their work in the affairs of human beings and other life forms of the biosphere. The hypothesis that we have considered in this book is that contingency is *real*; that is, it occurs outside the biosphere in the nonorganic world. It surely occurs in human affairs and is virtually the central feature of the human drama, which novelists, poets, dramatists, and cinematographers have long explored. It is clear that truth may at times be more bizarre than fiction or fantasy. I have maintained that contingency is indeed a *generic trait of nature.*

It surely is intrinsic to the human condition. Steve Allen, the famous

American TV personality and author, was driving home on October 30, 2000, bringing with him the galleys of his latest book *Vulgarians at the Gate*,[4] which was on the seat by his side. Meanwhile, a car suddenly veered into the path of his vehicle and struck it. This caused Allen to lunge forward and stop hard, his seat belt restraining him from going through the windshield. The man who struck him got out of his car, recognized Steve Allen, apologized, and asked for his autograph. Steve Allen smiled wryly in his typical way and responded, "Wow, that is the best attempt yet to get my autograph!" He proceeded home, told Jayne Meadows, his wife, that he did not feel well, and took a nap. An hour later, Jayne tried to awaken him and could not. Frightened, she called 911 for help. He could not be revived. An autopsy performed later indicated that he died of an aneurysm in his heart, most likely caused by the impact. There are causal processes at work here. Two motorists have an accident and the laws of physics apply: the rate of acceleration of one vehicle hitting another. There is also a biomedical explanation of his death.

Thus, there are two intertwined causal sequences: (a) the laws of mechanics enable physicists to estimate the acceleration and force of the impact; (b) the biological science of cardiology is able to locate the place of the aneurysm in the heart and postulate that it was most likely caused by the sudden pressure of the seat belt. There are also, of course, human explanations. The offending motorist was not attentive enough to guard against hitting Steve Allen's car; nor did Allen see him coming. These are generally known as intentional explanations. Here we need to move on to the ethical level in order to interpret behavior and perhaps affix blame. Were the actions of the motorist intentional or unintentional? This is important in the social context in order to ascertain liability by the motorist and his insurance company.

Here there are causal sequences at various levels, drawn from different kinds of explanations: physical, biological, psychological, and motivational; and these sequences interact at one moment in time, converge and collide. Perhaps it was an improbable event, unexpected surely, a freak accident. Was it due to *real* chance? There are many factors that are relevant in interpreting events in the context of human behavior. I have called this general mode of explanation *coduction*, since in any complete explanation we *coduce*, or combine, many kinds of explanations from many levels of analysis, and we do so all the time.[5] So a single cause may not be sufficient by itself to provide a full explanation. The tsunami, in which thousands were killed, can be explained by geological causes—an earthquake and a tidal wave. But the subsequent catastrophic mass drowning was also due to sociological and economic causes, which motivated people to build homes and hotels on the beaches. It was also due—in a negative sense—to the failure to establish a warning system and the lack of public education about the dangers that such tidal waves could have to life and limb.

There are abundant illustrations of coductive explanations that apply to historical events. As we have seen, the emergence of Nazism in Germany in the 1920s and 1930s was due to many political, economic, psychological, and sociological causes: the Versailles Treaty, which exacted what many Germans thought were excessive reparations; the resentment built up because of defeat in World War I; the desire for revenge; the fear among many industrialists of communism and socialism; runaway inflation; and so forth. A key factor was also the appearance on the scene of Adolf Hitler, an impassioned orator who shared all of these sentiments. And added to the goulash was virulent anti-Semitism.

Noted American philosopher and social critic Sidney Hook has written about the *Hero in History*,[6] pointing out that many decisive historical events were due to charismatic individuals who used the power of the state to fulfill their ambitious, sometimes idiosyncratic goals, rather than attributing them to underlying historical causes and trends.

In an ultimate sense, the role of chance, personal miscalculations, and psychological factors may be crucial. Thus Alexander, Napoleon, Lloyd George, Hitler, Stalin, de Gaulle, Kennedy, and other powerful leaders played causative roles, and we can always reconstruct historical events and ask, "What if particular historical events had not happened?"

We can repeat the exercise: What if the French fleet under Comte de Grasse during the American War of Independence had not blockaded Chesapeake Bay in 1781, which led to the defeat of the British Army and the surrender of General Charles Cornwallis at Yorktown? Would defeat of the American Continental Army at Yorktown have precluded the founding of the United States?

What if the French had decided not to defend Indochina in 1945? Would this have helped France to remain in Algeria, and would it have allowed the United States to avoid the costly Vietnam War?

What if President Kennedy had not been assassinated by a disgruntled malcontent named Oswald? Would the Vietnam War not have been escalated? This tragic war killed 58,000 American GIs and two to three million Vietnamese.

What if Joseph Stalin had not seized power as general secretary of the Communist Party in the Soviet Union? Would the tyranny that developed not have led to the near-total discrediting of the Russian Revolution of 1917? If Lenin had not died (or been poisoned), would Stalin have been thwarted in his ambitions?

Here is a line of hypothetical questions relevant to the late Cold War period: What if Mikhail Gorbachev had sent Russian troops to quell the uprising in East Germany in 1989? Would this have forestalled the collapse of the Soviet empire?

What if George W. Bush had not invaded Iraq for a second time in a debilitating war? Would an estimated fifty thousand US casualties, hundreds of thousands of deaths of innocent Iraqi citizens, and millions of refugees have been avoided?

Accordingly, human affairs are contingent—often dependent on singular events or personalities—and there are thus plausible countervailing hypothetical explanations. It is clear that human history involves not only economic, political, social, religious, scientific, and intellectual influences and trends in history but also *unique* events of real individuals, obdurate historical processes, and recalcitrant facts. So there exists *brute facticity* in human affairs, and *que sera sera* is *not* binding: "What will be, will be" is *not* necessarily the case, for it also depends on what human beings do. And the role of other unexpected singular events, odd personalities, strange quirks, and bizarre happenings may not have led to what eventuated. All too often bizarre happenings intervene: the *Ripley's Believe It or Not!* series applies so graphically to history and life.

An especially dazzling illustration of this is the role of *scientific discovery* and the unpredictability of technological inventions and innovations that issue from it. Thus there are "intellectual" causes at work in human history. Marx's economic (or sociological) interpretation of history emphasizes the central role that the forces and relationships of production in the base play as the decisive causal factor in social change. He places in the superstructure religious, political, moral, intellectual, and cultural factors. He said that it was the *economic* factor that was most pivotal. Yet scientific research, which is dependent on intellectual discovery, is surely a "force of production" and indeed along with technological innovation should be part of the base.

There are some telling illustrations of the role of scientific discovery in technological and economic changes: What if Alexander Fleming in 1928 had not accidentally discovered penicillin while working on a mold of staphylococcus bacteria and observing a bacteria-free circle? The cure of infectious diseases might not have been developed until much later, with many more people dying in the interim.

What if Galileo had not made his discoveries in astronomy and physics? The science of mechanics, which subsequently led to changed principles of warfare and the Industrial Revolution, might not have developed until much later.

What if Nobel Prize–winner Herbert Hauptman had not codiscovered the structure of crystals in 1985 on which a significant part of the pharmacological industry has based so many effective medicines and therapies?

What if Darwin had not gone to the Galapagos to lay the foundations of the theory of natural selection? Well, you may respond, it can be argued that Alfred Wallace had also discovered it. Yes, but it might not have had the profound impact that it had, especially if a remark by Bishop Wilberforce had not mocked evolution and if T. H. Huxley, the bulldog defender of Darwinian theory, had not put him in his place. This had a major impact on our understanding of human nature and has helped to undermine religious doctrines of creationism and intelligent design.

Perhaps more pointedly: What if Karl Marx had not written the *Communist Manifesto* and *Das Kapital* and other tomes found in the British Museum? Would communism not have developed the way it did?

And what of the later development of powerful religious institutions? What if the camel herder Mohammed (who perhaps was suffering from depression and hallucinations) had not considered his "visions" of the angel Gabriel as genuine revelations transmitted from Allah, and had not gone on to raise an army to conquer Arabia and thus fulfill divine commandments? If not for Mohammed's subjective experiences, Islam might not have developed as a world religion.

What if Jesus had not believed that he had a special mission and had not died on the cross (perhaps he never lived in the first place), and what if Saul of Tarsus had not thought that he had a message from Jesus on the road to Damascus (was it due to an epileptic seizure?) and had thus not become St. Paul, the key founder of Christianity?

As can be seen, contingent historical events often lead to totally unanticipated consequences, much to the surprise of everyone, and after-the-fact reconstruction of origins leads to a second astonishment at the powerful developments that simple beginnings may, over time, accrue. It is difficult to predict where a human invention, discovery, or random event will eventually lead. *Que sera sera* will not necessarily be; it depends on accidental historical facts and chance at any one moment in time.

FREEDOM OF CHOICE

Now, I realize that an enormous amount of effort historically has been expended on the free will versus determinism debate. In the past I have taken the position of the soft (rather than hard) determinist, arguing that the reality of *both* causal conditions *and* choice in human affairs may be present at the same time. I submit that the problem has been misstated by the hard determinists who suppose that there are *hidden unknown causes* in human choice. First, the supposition of the theist, that the mind of God is the underlying basis of causality in history, is mistaken. This is surely a speculative leap of faith: God is the unknown entity invoked to *fill the gap* by supposing that there *must be* an ultimate, though unknown, cause. The fallacy in this line of reasoning is that it is based on a pure postulate, sustained as an article of faith. But there is no evidence that a *deus ex machina* explains anything because it is simply thrown in to account for phenomena; and occult causes had been rejected long ago by scientists since they provide no explanation at all. This is the pious sigh of those unable to account for what happens, and it is an obstacle to genuine scientific inquiry itself, which could not proceed on such a mystical foundation. The first natural philosophers (Galileo, Newton, Kepler, et al.) realized that introducing God to fill the gap is fatuous nonsense. There is no justification for the injection of such an untestable claim.

But there is a similar fallacy introduced by so-called scientific determinists, who are skeptics about the "God of the gap" explanation, though they are predisposed to believe that there *must* be hidden physical causes. Ultimate physicalist causes *may* be at work, perhaps in the neurological or physical-chemical operations of the brain, which, although presently unknown and unverified, exert a causal influence. Emergent choices by humans, however, are *not* admissible for these determinists. This is surely an *a priori* supposition, which is based on faith, not evidence. It predisposes fundamental scientific laws to be the base of all events; it again supposes that the universe (if *not* God) "would not play dice" with human motivation. But scientific determinism is only a presupposition. Clearly, there are concomitant firings of neurological networks that *accompany* decision-making behavior, but this does not preclude the possibility that there may be other forms of behavior that function at the same time at different levels of interaction.

Stop! I ask the reader to consider this question: As you read what I have written, are you prepared to change your beliefs or hypotheses about how cognitive processes work, or at the very least to be influenced by arguments on the level of intellectual debate, and/or the introduction of evidence on the level of scientific inquiry? Do we not have an emergent property within an argument and inquiry on the level of human discourse that cannot be reduced simply by the suspicion that there must be hidden and ultimate causes for all psychological behavior—including *the process of argument and inquiry*? This surely cannot justify the deterministic case "to fill the gap." An argument is an argument; it is not simply a set of neurological firings, and a scientific (or philosophical) inquiry is an inquiry that goes on involving persons, writings, publications, discussions, and controversies, which cannot be reduced *ipso facto* to the physicalist atomic hypothesis (powerful as that is); for these are epistemological criteria, principles or inferences, standards of confirmation and corroboration, which have their own internal integrity and rationale, as understood by logicians, scientists, and ordinary people of common sense. And the attempt to explain them away by insisting that there *must* be some underlying cause(s) for human choice and that a person does not make up his mind freely is sheer nonsense.

Indeed, the typical way that the question has been framed is illegitimate; that is, attributing choice to "the will," whatever that means, and saying that it is contra-causal or "free," which is not clearly defined. Nor would I attribute choices to the "mind," a quasi-mystical notion based on a mind-body dualism. Moreover, our cognitive choices—especially about personal matters—are accompanied by emotive attitudes.

I readily concede that choices are caused and conditioned by a whole range of contingencies, but nonetheless it is the person who makes a choice, and this act of choosing is a form of *behavior* of a human being, including his or her body, brain, and nervous system, which functions in a sociocultural and natural environment. Hence I can say, *I made that choice* or *you made a choice*; indeed, we can both be held responsible for our choices, and in many cases you and I can be persuaded or convinced to change our choices in the light of reasons or evidence on the level of cognitive and emotive behavior.

I voted for Barack Obama in the 2008 presidential election, though I originally supported Hillary Clinton. And I listened carefully to the arguments and weighed the reasons why I should. In the past I often changed my beliefs and attitudes (as did you, dear reader), and I am willing to change them again in the future, if need be.

Human beings do make choices in the role of practical decision making in human affairs, selecting candidates or platforms, and deciding to buy or sell products and so forth, listening to people, agreeing or disagreeing with them, and this is the very fabric of social interactions, and it presupposes a rich cultural tapestry, which has influenced me and which I can accept or reject.

Let me raise this question: Can human beings learn from experience, can they change the way they behave (within limits), or are they blind automata, simply responding to hidden stimuli and acting in the light of conditioned responses?

Let us move to a test criterion and ask whether you "could have acted otherwise" in any of the decisions that you made in the past. Or were these decisions forced upon you? Were you compelled to do what you did?

"I COULD HAVE ACTED OTHERWISE"

We can and do speculate about "what ifs" *retrospectively* and what might have happened if we acted differently. We cannot, of course, undo the past, though we may choose to ignore it, but we are constantly blamed or praised for what we did. Similar considerations apply to the choices that we will make prospectively about the future. Although a person is a product of a wide range of influences and causes in his past, his personality and behavior is the result, in part at least, of the choices that were made and the historical events that influenced him at the time. The person that you decided to marry or divorce, or the career choices and job offers that you accepted or rejected may have been due to accidental circumstances, luck, misfortune, or limited options, but you had a hand in the choices. The course of study that you decided to undertake in school or college, the person you fell in love with, the partner (husband, wife, or lover) you chose, are to some extent within your power, in the sense that you could have chosen not to; and many people constantly berate themselves for imprudent choices and resolve to improve in the future. We may speculate about everyday choices that may have serious consequences that we later regret: "I should not have given in to my sexual passions when I did, but I couldn't resist the temptation."

"I can give up smoking whenever I choose to; I have done so a hundred times."

"I should not have had that glass of scotch; I know that I can't handle liquor. I should not have driven my car home inebriated."

"I should not have majored in law; I find it tiresome."

There are profoundly serious dilemmas facing individuals about questions of life and death, career choices, marriage and family. There are also personal disagreements with friends and relatives that may require sacrifice and involve heart-rending decisions. Similar considerations may apply on the social level to the community, school, company, region, or national state. Political, economic, or military decisions proposed may be life-threatening. We may face crucial turning points with massive consequences, such as the decision to go to war, seek victory, or accept surrender.

What we decide to do as individuals depends on our propensities, habits, and predilections, and these are often so deeply ingrained that it's sometimes difficult to resist them or act to the contrary. Who or what we are is a result of a wide range of social and environmental forces that have conditioned us. Moreover, a person's genetic tendencies and ingrained psychological passions may be so powerful that it often requires an insuperable effort to resist them. Our choices are constrained or impelled by these basic causes. All of this needs to be understood as granted. If a person is gay or a lesbian (a genetic tendency), there is very little, given what we today know about sexual propensities, that can be done to change one's nature, though sexual behavior surely can be repressed or controlled to some extent. If a person has a low IQ or lacks musical talent, there are certain professions that he or she could not enter with much hope of success. Many choices that we make may depend upon our deep-seated preferences and desires; thus it may be difficult to resist our deep-seated tendencies and tastes, and insuperable efforts may be needed to resist them; we may be facing overwhelming temptations, as we are all too human.

On the other hand, human beings often are able to act contrary to their deep-seated psycho-bio-social personality traits and proclivities. Reason can at times redirect attitudes and modify emotions. Thus, human beings can resolve to do or not do something, even though underlying impulses and motives may be difficult to resist. We can respond to arguments and be persuaded by reasons, and we can act contrary to what is normally expected. We are, I submit, capable of voluntary choices in spite of the above, and we can act responsibly. We are not blind automata. We can master our own destiny, provided that we do not live in a rigid society that constrains freedom of choice.

Where there is choice, there is still some measure of freedom. *Soft* as distinct from *hard* determinism affirms that the behaviors of human beings, including their decisions, are conditioned by a wide range of causal factors on many levels—genetic, biological, psychogenic and sociogenic—yet freedom of choice is a creative dimension of human behavior, and it can and does add something new to the human equation; not in an absolute but in a relative sense.

My inner beliefs and convictions are within my power to some extent (we always need to add a qualifier), and I can change them in the light of new evidence and reasons. Indeed, in some sense, who we are is the result of plans and projects that we have conceived and the motivations that we have summoned to achieve them. The wish is father to the fact; our goals are products of our desires and purposes, and cognition can control or moderate the things that I may like or dislike and create new ones. Human behavior often involves taking chances and undertaking risks, gambling on our hunches and resolving to bring them about. Some individuals are fearful of these opportunities; others may say, "What the hell" and leap in. Hard determinism, I submit, is akin to a religious faith, worshipping at the altar of hidden causes. It is contrary to who and what we are as resolute and responsible human beings. It is also a false doctrine often promoted by malevolent fools who may wish to control other people's behavior.

In one sense the strength of determinism or freedom depends on the society in which we live and the kind of personality we have developed. Highly authoritarian societies expect conformity and make it very difficult for individuals to act against social mores. Open libertarian societies, on the other hand, encourage ingenuity, individual initiative, entrepreneurial adventure, exploration and discovery, innovation, and experimentation, and above all free thought. And individuals *are*, I submit, capable, at least on some occasions, of resolute action.

Thus I would argue that free societies are more likely to encourage libertarian behavior and freedom of choice and are more likely to allow individuals to excel on their own terms. Tiger Woods, the outstanding golfer, had great training in his life, yet he exerts every effort and strains to exceed all others in the competitive matches that he plays. (Though he has had several torrid love affairs, which shows that passion is a powerful influence on behaviors, they have not affected his performance on the golf course.) Similarly for Michael Phelps, who exerted every fiber and muscle to break records in the Olympic swimming competition. Yes, we can succeed, if we have the fortitude to do so; at least many individuals believe that and act upon it.

Lazy, indifferent, conforming individuals tell us what cannot be done: creative, life-affirming, heroic individuals express the audacity to achieve what they wish, as far as they can. Nothing ventured, nothing gained. Freedom of choice, in one sense, defines who and what we are as persons on the cutting edge of life. The great scientists, artists, composers, philosophers, statesmen, captains of industry, and builders of new futures look ahead, embark on uncharted seas, and are ready to exert bold efforts to realize their goals. They exude exuberance. And so do ordinary men and women in their own domains.

Thus *moral choice* is a key ingredient in a contingent universe, for what will be is not predetermined but rather is *post-determined* by what human beings *determine for themselves what they will do*. And this depends on freedom of inquiry, the discovery of new truths, artistic creativity, moral inquiry, and resolve. It is an emergent quality that has defined human civilization, a Promethean leap ahead, the courage to *become* who and what we are, at least in some sense: surely in regard to whom we make love to or marry, how we bring up our children, or how we choose a job or career. Of course *chance* plays a role, and who we meet and what we do depends on our opportunities and options, which may be wide open or limited; so *luck* is a factor, being at the right place at the right time.

Similarly on another plane, many collective decisions draw on the traditions, customs, or laws of a nation, the competing interests and rivalries and the cooperative opportunities that are available. Many people believe that societies are difficult to change and that reform is often impossible to effect. This allegedly constrains our freedom to act, without a violent revolution or war or total collapse. Be that as it may, does this preclude any and all sociopolitical movements from changing a society? I think not, for the historical record demonstrates that even the most authoritarian societies have been overthrown and that unexpected historical contingencies have undermined ancient regimes and totalitarian systems, which may eventually capitulate if the people rise up in protest.

The question of determinism versus free will depends on one's conception of causality. To say that "X causes Y" means that "whenever X occurs, Y is a likely effect." In time we develop habits of expectation and come to depend on them. Thus our world is orderly; these *are* regularities that we can count on. The four seasons, if we live in a temperate climate, allow us to plant seeds in the spring, water them in the summer, and harvest them in the autumn, unless there is a heavy drought or flood that destroys our crops. Bees help to pollinate flowers; the honey cultivated in the combs is edible and sweet. The sap flowing in maple trees in the spring provides delicious syrup. If the harvest is plentiful, we can sell our crop or store it in silos for winter feed.

If I launch a rocket into space and if I know its direction and velocity, I can with precision calculate its path. In a total eclipse, the Moon will cover the Sun and block out light from it. By telescopic observations I can plot with precision the orbit of a comet. I can boil water at 212° Fahrenheit and freeze it at 32° F.

A body of knowledge is thus built upon that which we can comfortably rely. There is a kind of order that we can discern and count upon. Given certain causal conditions, all things being equal, certain observed effects will most likely ensue. Armed with knowledge of infectious disease, doctors can diagnose the cause and prescribe medication to control it. If a person has a tumor, it can be removed by surgery if it is not malignant; if it is, special precautions need to be taken. Knowing something about the stress of materials, engineers can construct bridges, bore tunnels, or erect skyscrapers to withstand collapse.

The question is often raised, if we know the condition under which certain effects are likely to occur, could we in time develop a body of theories and laws that would enable us to predict with a degree of accuracy what is likely to ensue anywhere and everywhere? Mathematics enables us to develop theoretical principles with precision and comprehensiveness. Can science provide us with sufficiently reliable knowledge to succeed in all of the endeavors we undertake? Scientific knowledge has developed rapidly in the past few centuries in the natural sciences (physics, astronomy, geology, chemistry), the biosciences (biology, genetics, etc.), and the social sciences (sociology, economics, anthropology, psychology, politics, etc.).

Methodological naturalism has developed a set of methods and strategies for testing hypotheses and developing theories that are reliable. There are certain rules governing scientific inquiry: (a) We should look for natural causes, eschewing occult explanations; that is, causes that are amenable to intersubjective observation by the community of objective inquirers and/or that can be confirmed experimentally by their predicted effects. Replication and peer review are essential in any domain under investigation. Thus there is empirical/experimental/evidential corroboration. (b) Theoretical constructs can be introduced using mathematical extrapolation grounded in experimental data. These analytic tools are validated by logical inference, and they are modified in the light of new evidence or the introduction of more comprehensive explanations. (c) Scientific methodology is reliable but not infallible, and scientific inquirers are prepared to modify their hypotheses and theories in the light of new evidence or more powerful theories based on mathematical precision and comprehensiveness. Thus science is fallible, skeptical, probabilistic, ever seeking more reliable explanations. The role of contingency in nature accompanies fallibilism in human knowledge, because even the most airtight theories have been undone by the uncovering of new facts—the "damned facts," by observing unanticipated eventualities and by introducing new hypotheses and theories that provide more effective explanations.

The one implication that I wish to draw from this discussion is that if the world—our world—is contingent, then we have a role to play, and I (or we) can intervene in the natural order of things and affect our environment of interaction. Fatalism is mistaken, whether our fate is attributed to the "hidden hand of God" or to the underlying "causal network." Thus our decisions and the behavior that ensues from them are causal. Our intentions, which lead to actions, can affect what happens and change things. This is sometimes known as motive-explanations or intentional causality. Given this rule for freedom of choice, our future destiny is not fixed, and I can and do have something to say about my future; similarly for collective decisions made from within by companies, universities, voluntary organizations, political and economic institutions. Whatever will be—in part at least—depends on what I (or we) choose to do.

Accordingly, we are the masters of our destiny, at least partially so. That is why the formulating of long-range projects and plans is essential if we are to succeed. We need goals, blueprints, and designs of what we wish to bring into the world; hopefully such planning can be rationally formulated. We need strategic planning, yes, but we also need to be aware of the best short-range tactics to realize our long-range purposes and goals. Tactics refer to the means (i.e., our resources, instruments, technological tools) that we have to use. But we also need to evaluate them by *clarity* as to our ends and purposes. Actually there is no sharp distinction between the importance of means and ends; we need to be aware of our goals *and* the means that can be adopted to realize them. Accordingly, there is a means-end continuum, and the means at our disposal or that we need to bring to bear or invent may modify our ends and vice versa. Both should be taken into consideration—in a cost/efficacy evaluation. Targets may need to be readjusted in the light of empirical inquiry. Both the cost *and* the consequences of our means in the world of affairs need to be examined honestly, because the means may be so destructive of other values that we may hesitate to apply them. The *risks* also need to be part of the appraisal process. The point is that human beings are capable of intelligent choices made after a process of deliberation. And although not all problems can be easily solved, we nonetheless can apply rational considerations to the decision-making process. In many cases we could have acted otherwise, and we learn from trial and error that some choices are indeed better than others. Our decisions are relevant to the concrete situations in which we make them.

Although there may be general guidelines that apply in situations—like the ones we now face—what to do depends upon the circumstances that confront us and the genuine options that we have. This emphasizes the contextual nature of the decisions that we make. No doubt there are general *prima facie* rules that are relevant in similar situations, but how they are evaluated often depends on the unique facts of the case; the person or persons involved; the consequences of alternative courses of action; the means at our disposal; our preexisting attitudes and values, desires, and satisfactions. Our decisions may be weighed on a comparative scale of better or worse. We balance competing factors in order to choose what seems most fitting in the light of a wide range of considerations. Our choices are very rarely absolute or categorical. Nonetheless they need not be capricious or subjective, and although they are relative to who is involved, where it occurs, and when, they may in some sense be *objective*, especially if they are reached after a reflective process of inquiry. In many cases in life, *we could indeed have acted otherwise!* However, I agree that in some situations choice may be very difficult and predilections may be so powerful that it may be very hard or virtually impossible to change what we will do.

Nevertheless, some confidence in our own capacity to change or redirect the course of events is essential if we are to get through life; some optimism about the likelihood of success is a vital ingredient of pragmatic coping intelligence, which, after all is said and done, is the only thing that we can rely upon in a constantly changing world.

ETHICAL QUANDARIES

What do I conclude from this? That often in life there is no perfect solution and that we do the best we can, drawing upon both reason *and* passion. However, I submit that there are some guidelines in the best of circumstances. In some cases, the situation may be hopeless. It all depends upon the level of knowledge attained and whether a compassionate good will is present.

There are, of course, awesome moral quandaries that humans face that are difficult to resolve. Here are some examples:

1. A married woman with two children falls passionately in love with a married man with children. Should they get divorced and remarry? Are there any moral principles to guide them?
2. A relative is totally paralyzed and in great pain. He pleads with you to assist him in committing suicide (a form of active euthanasia). Is assisted suicide ever warranted? If so, under what conditions?
3. The company that I work for deceives customers and sells them shoddy products. Should I report them to the district attorney and possibly lose my job as a result?
4. There are also moral questions that confront social institutions. My country—right or wrong—is engaged in a secret war to wipe out insurgents, though in reality many of them are innocent victims. Shall I refuse to obey my commanding officer's orders? Is this ever justified?
5. President Harry Truman ordered the bombing of two open Japanese cities, Hiroshima and Nagasaki, which killed two hundred thousand civilians, in order to shorten World War II and save the lives of American GIs. Was this justified?
6. Vice President Dick Cheney (from the George W. Bush administration) condones the torture of enemy combatants in order to extract information from them about possible future terrorist attacks. He said that this was justified to protect innocent civilians. Is this morally permissible?

IN RESPONSE

None of the problems have easy solutions. It is often the lesser of two evils.

- Response to (1): The problem of separation and divorce is widespread in society, and the individuals involved have to weigh alternatives carefully, including the effect upon the children. There is no optimal solution; it's simply a question of balancing goods and bads, rights and wrongs. Though monogamy is widely praised in many societies, it is the exception rather than the rule. Any choice made must consider all the parties concerned, including the children.

- Response to (2): "Mercy killing" used to be a tragic choice, especially in strict religious societies that consider euthanasia sinful and absolutely forbidden. Many nations still consider this murder. Today, some societies have loosened their opposition, and there are conditions under which it is justified on the basis of compassion and reason: If the individual who requests assistance is dying and suffering intractable pain and there is no significant quality of life, humane societies will permit active euthanasia if two doctors testify to the person's condition and if there is a court order guarding the person's right against compulsion from greedy kin.
- Response to (3): Employees should have the right to report egregious practices of a company. However, he or she should be protected from retaliation under a provision of law that allows "whistle-blowing."
- Response to (4): The conscience of humankind has progressed concerning how wars are to be conducted and which acts are impermissible in military behavior. Soldiers in the field are not required to commit heinous acts, even if commanded to do so. The intent of the order should be to advance against the enemy and protect soldiers in the field. All efforts should be exerted not to purposefully harm innocent civilians.
- Response to (5): The decision of President Truman to totally destroy Hiroshima and Nagasaki, with tremendous civilian casualties, is open to dispute. American patriots claim that Truman was right to shorten the war and thus protect the lives of American soldiers and sailors. Critics say that the massive killing of innocent civilians is morally impermissible. Defenders of Truman say that the intensive bombing of open German cities by British and American planes was total war and was necessary to defeat Hitler, not unlike the bombing of the Japanese cities. It was retaliation for Nazi war crimes. Critics object, saying that neither of these acts of destruction of civilian populations was morally justifiable and that it is necessary wherever possible to humanize how wars are conducted.
- Response to (6): The Geneva Convention clearly states that the torture of prisoners of war is morally impermissible. In response, the defender of torture (such as Cheney) poses the catch-22 question "If a terrorist knows that a nuclear bomb will blow up Paris or New York, is it not permissible?" There may be an ultimate exception if a much greater evil

will ensue by not extracting such information from prisoners of war. Critics respond that the exception does not justify the torture, and that abiding by the *prima facie* general principles against torture is an effort to humanize the rules of war.

CONCLUSION

Ethical problems are often difficult to solve; that is why so many are called *dilemmas*. It is clear that in such situations we need to clarify the facts of the case. In judging what to do, we need to weigh the alternative courses of action that can or could have been selected (if after the fact). In some situations it may be a question of choosing the lesser of two evils. Our decisions are comparative, in which we understand the options that we have, the means at our disposal, and the consequences of various choices. We need to take into account the values that we cherish and the moral (and/or legal) principles that we believe we are constrained to follow. In all such cases, our decision is reached after weighing the facts, means, consequences, values, and norms that we think are relevant, and our choice is that which on balance seems most fitting.

ACT NINE

ORDER AND HARMONY IN THE UNIVERSE

BONDING

There is disequilibrium and chaos throughout the turbulent universe, typified by the groan of a dying man in an automobile accident, the squeal of a pig about to be slaughtered, the moan of a mother in premature child labor, and more—a raging "perfect storm" overturning everything in its path, the explosion of a supernova in outer space, the collision of two galaxies.

No doubt the above characterization is overly pessimistic, for we also find in the universe unity, harmony, balance—and especially bonding. The attraction of electrons and photons; the gravitational forces that keep planets swirling around stars for eons of time; the cohesion of millions of suns in a galaxy adhering and circling around an invisible center; electromagnetic forces; the homeostasis in organisms that maintain internal and largely unconscious equilibrium; the union of a female and male in erotic passion; the enduring shared affection within a family; the bonds of male warriors or athletes in cooperative teamwork; the uniting of factions overcoming differences; the negotiation of a new basis for cooperative action and unity, permitting institutions to emerge in uplifting harmony.

Indeed, without some order it would not have been possible for life to have evolved on Earth, for the timescale is slow enough for forms of life to survive and reproduce. Although the Earth is hurtling through space at a dizzying rate of speed, we are unaware of it; the "original state of nature" is peaceful and quiet for much of the time. Thus, flowering plants, grazing animals, flocks of geese, and human societies can persist because of it. Yes, there are places of silence and solitude, such as the forests in the Canadian wild where the overwhelming sense is quietude. I remember well my visit to the African nature parks. Sitting in an enclosed truck in Kruger Park (between South Africa and Mozambique), we were astonished when an elephant silently appeared; we barely heard a twitch until he was next to us. The overwhelming presence was an all-pervading silence. At times we heard birds at a distance and at night the sounds of distant hyenas. There, on top of the mountain in Mouans-Sartoux, we heard the grunts of frogs and in the hot weather the clickity-clicks of *criquets*.

In particular, the fact that humans have survived was due, as I indicated, to their wily coping skills, but even more to the evolution of culture. The fact that I have emphasized the pervasive conflicts in human affairs between competing marauding tribes, ethnicities, and nations does not deny that side by side with this was the gradual evolution of systems of law and morality. The Roman Empire survived as long as it did because it established a body of law, a network of roads and water reservoirs, and legions of soldiers to protect the borders from the barbarians and the urban centers from unrest and uprisings. Moreover, the gradual evolution of religion sanctified a set of moral commandments, guaranteed peace and civility, but also educated the children and instilled in the masses the need for obedience to moral principles: "Be righteous; do not envy thy neighbor," "assist the alien in thy midst," "love one another, even as I have loved thee," and more. These were enforced by the divine sanctions of heaven and hell. Similarly, the great thinkers of the past tried to define *good*, *right*, *justice*, and other moral ideas; for the Platonists there was an eternal realm of essences. Other philosophers, such as the Aristotelians, thought that moral rules had a kind of objectivity that rational persons could recognize.

HUMAN RIGHTS

There has been a slow development of ethical ideals in human history, a process that has been progressing in the modern world. This is especially the case with the emergence of democratic institutions, ignited by the American and French Revolutions, though this can be traced back to the English Magna Carta, delimiting the rights and privileges of the people against tyrannical monarchs.

The Magna Carta was an important document, for it declared a supreme law of the land. King John of England agreed to it in 1215 in response to the demands of his barons. It stated that

> no free man shall be taken, imprisoned . . . or in any way destroyed . . . except by the lawful judgment of his peers or by the law of the land. To no one will we sell, to none will we deny or delay, right or justice.

The king affixed his seal to the Magna Carta and read it out in public, and he authorized handwritten copies to be made. By doing so, he bound himself and his heirs in the future. It was important, for it affirmed that the law must be above the king, and even he could not break it.

There are other documents that stand out today as forerunners of the recognition of human rights. They are intent in securing the rights of citizens to establish democratic representative republics with limited powers.

The Declaration of Independence was important. It reads as follows:

> The unanimous Declaration of the Thirteen United States of America, in Congress, July 4, 1776.
>
> We hold these truths to be self-evident, that all men are created equal, that they are endowed by their Creator with certain unalienable Rights, that among these are life, liberty and the pursuit of Happiness. That to secure these rights, Governments are instituted among Men, deriving their just powers from the consent of the governed.

This was drafted by Thomas Jefferson and redrafted by those present at the Continental Congress. The Constitution of the United States was later ratified by the new states in 1789, and subsequently ten amendments were adopted as the Bill of Rights, recognizing other rights, including freedom of the press, the right of assembly, and the right of the people to petition for the redress of grievances. It recognized freedom of conscience and religious liberty, and it prohibited the establishment of religion—these were all foundations of the secular republic.

There is no evidence for the notion that rights "are endowed by the creator." They are *human* rights and they are justified empirically, for to live in a society that does not recognize the rights that have been fought for in human civilization may lead to tyranny. The doctrine of human rights may be justified independently of religion. All too many people who have believed in a creator have rejected the concept of rights. A right is a *moral ideal* that humans agree to respect for a variety of motives and reasons.

Another great document defending human rights was issued in France, as the Declaration of the Rights of Man and Citizens by the National Assembly following the French Revolution of 1789. The first right is as follows:

1. Men are born and remain free and equal in rights. Social distinctions may be based only on common utility.
2. The aim of all political association is the preservation of the natural and imprescriptible rights of man. These rights are liberty, property, security, and resistance to oppression.
3. The principle of all sovereignty resides essentially in the nation. No body nor individual may exercise any authority which does not proceed directly from the nation.

Another key affirmation came a century and a half later. It was the Universal Declaration of Human Rights of 1948, issued by the General Assembly of the United Nations. Its preamble begins as follows:

> Whereas recognition of the inherent dignity and of the equal and inalienable rights of all members of the human family is the foundation of freedom, justice and peace in the world. . . .

Many nations of the world have become independent by popular uprisings, revolutions, or wars of liberation, and were inspired by similar sentiments. Unfortunately many regions of the world that are dominated by religion (such as Islam) do not accept them. Today a new consensus about democracy and human rights has emerged. The growth of democracy has by now become widespread. It is grounded in a commitment to humanist values. It provides a solid foundation for the future of humankind. These values are still being expanded. They provide a common foundation for the civilized conscience of the planetary community.

Among these rights are

- *The abolition of slavery* in the nineteenth century by the United Kingdom, France, Spain, and eventually the United States.
- *The suffragette and feminist movement*, an ongoing battle in the nineteenth and twentieth centuries to extend full and equal rights to women, including the right to vote.
- *The Geneva Conventions*, enacted after World War II, developed certain prohibitions and duties of nations about the rules of war, such as not inflicting casualties on innocent civilians and the humane treatment of prisoners of war.
- *The enactment of social welfare programs*, providing reasonable retirement funds for elderly persons, usually based in part on employee contributions.
- *Universal healthcare*, the recognition of the right to health of every person within a nation, whether by private or public organizations. No person shall be denied healthcare.
- *The right of workers to organize* and engage in collective bargaining with their employers over wages, hours, and other benefits. This has led to the development of trade unions worldwide. It has contributed to increased standards of living and job security.
- *Unemployment insurance* recognizes the right of everyone to be gainfully employed, wherever possible, and if a person loses a job, the right to some compensation in the interim.
- *The rights of the child*, among which are the universal right to education, providing balanced knowledge of science and the arts; the right of the child not to be harmed, excessively punished, or forced to engage in child labor; the right of the child and young adolescent to freedom of conscience and religious belief or the lack of it, the right not to believe; the right of the child to adequate food, healthcare, shelter, and safety.
- *The rights of homosexuals and transgendered persons.* All persons shall be afforded the same rights and privileges as all others, including the right to cohabit or marry, or to raise or adopt children, regardless of their sexual orientation.

There are other rights that may be recognized in the future. Many nations dominated by religious oligarchies refuse to accept human rights, especially those of women, homosexuals, and dissenting religious minorities.

DEMOCRATIC REVOLUTIONS

Human rights became a political rallying cry historically when democratic revolutions defended them. At this stage of human history democracy has emerged as the most popular form of government, especially when compared with other political systems—monarchy, aristocracy, oligarchy, and dictatorship. Political democracy best flourishes in a democratic society in which there are a multiplicity of voluntary associations and institutions. Such a society is one that attempts to maximize grassroots participation of the people in the affairs of the society, and it provides a wide range of freedoms, allowing citizens to make their own choices about personal matters.

A democratic *state* elects the key officials by majority vote. Representative government is most effective in an open society where freedom of speech, the press, and freedom of conscience are safeguarded. It protects the rights of dissenting individuals and minorities, and the legal right to oppose the ruling party.

Democracy presupposes a set of basic *ethical principles*. All citizens are equal before the law. The principles of justice apply to each person without special privilege or favor. It presupposes that the citizenry is well informed. This depends on universal education, which is open to the children of the poor as well as of the wealthy. For democracy to be effective, the public must be informed about the major decisions and policies of government enacted by the legislature (or Congress) and applied by the executive branch. And when disputes arise regarding these laws, an independent judiciary is on hand to interpret their meaning, extent, and application in an impartial manner.

Democracy operates on the premise that "he or she who wears the shoe best knows where it pinches." The open society is less liable to cruelty and duplicity, and unjust conditions can be scrutinized by the press and be reformed. Democratic government depends upon free and open discussions in which different points of view contend. It recognizes that we can learn to appreciate alternative viewpoints. This entails the willingness to negotiate differences, to work out compromises, to accept the laws adopted, and to agree to live by them. The *right to privacy* is widely accepted, and a distinction is made between the private and public domain. Therefore, the government has no right to enter the bedrooms of adults.

Political democracy by itself in some countries may be simply formal—unless the economic sphere is open to reform. A democracy is most effective where there is a large middle class. The society should thus be concerned with redressing economic deprivation and hardship. That is why minimum-wage laws, unemployment insurance, universal healthcare, and social security have been enacted in virtually all democratic countries. The concept of social welfare is widely recognized, though many economic libertarians may oppose it.

What is essential is *equality of opportunity* to improve the status of everyone in society. Universal education is crucial in allowing all individuals to go on to college or the university. Laws against racial, religious, ethnic, or gender discrimination are also a prerequisite. People need to be encouraged to aspire to new levels of excellence. Much of this has already been adopted in democratic societies. One way to achieve these goals is by means of progressive tax policies and by affirmative methods of instilling motivation in deprived individuals to rise to the challenge. As science and technology continue to contribute to economic and social productivity, opportunities need to be made available for changing jobs at midstream and for providing continuing education for all age groups to modify their values and capabilities in light of altered economic and political conditions. An individual cannot rest on his or her past laurels; thus the best guarantee of social mobility is to teach people how to adapt to changing circumstances. The most reliable guide is the development of *thinking skills* and the willingness to modify one's values in the light of inquiry. Coping with uncertainty is a prerequisite for leading the good life in contemporary society.

The question is often raised: Is what I have stated simply an expression of Western ideological bias? I think not, for I believe that democracy can be justified empirically for any country in the world. In the first half of the twentieth century there were widespread debates about the efficacy of various forms of government. Many socialists and communists indicted capitalism and sought to replace its economic system by a cooperative social system. Alas, the domination of the economic marketplace by socialist governments was often insufficient, so today there is widespread recognition of the dynamism of market forces. Virtually all communist countries today—such as China and Vietnam—have managed to leave room for markets. Mixed economic systems seem to be most efficient, combining both a private and a public sector; a free-market economy, and government regulation. Governments can invest in sectors where private interests may be reluctant to enter—though we need to encourage private initiative and investment.

In the 1920s and '30s fascists installed a corporate state that controlled the economy. The result was disastrous, for there was abuse of human rights, which were sacrificed to the war economy of the state. When Mussolini came to power in Italy many thinkers applauded him because of new efficiencies. "He got the trains to run on time," it was said. Yet without an open society, rampant abuses, cruelty, and corruption were the result.

How is democracy to be justified? I contend that it must be done on *pragmatic* grounds. Political and social democracy are preconditions of a just society, and unless there is an open society in which the democratic electorate has a key role to play in formulating public policy, there is no way to guard against the perversion of power by the state. Accordingly, the best guarantee of human rights is the right to criticize the policies of the government and overturn unjust decisions and laws. Democracies need leadership; however, the intelligent distrust of leaders is an essential component of the open society.

NEW TRANSNATIONAL INSTITUTIONS

I would add to the above other important planetary recommendations that I think are necessary for the future of humankind. Much of this is proposed in *Humanist Manifesto 2000*, which defends the development of new transnational institutions as prospective moral ideals.[1]

- *The Establishment of Transnational Institutions* and their continued growth are accordingly necessary. We are aware of the League of Nations, which, unfortunately, did not succeed. But it was replaced by the United Nations. Weak as that organization is, it nonetheless does important work. New planetary institutions need to be developed over and beyond the United Nations.
- *UNESCO* (United Nations Economic and Social Council) provides a basis for continued cooperation on many fronts, assisting economic and social reforms. Its programs need to be expanded manyfold.

- *A Planetary Court* to place on trial, prosecute, and, if the evidence warrants, convict suspected war criminals as well as persons who have resorted to heinous crimes that offend the conscience of humanity. A precedent for this was the Nuremberg trials of Nazi war criminals, but it is essential that this be extended to include criminal behavior in the major powers as well—such as perpetrators of preemptive wars. If human rights and democracy are to be extended, then a set of planetary institutions should be developed. There are two such courts with limited jurisdiction: the International Court of Justice (ICJ) and the International Criminal Court (ICC).[2]
- *World Law*. It is vital that the planetary community develop a transnational body of laws to govern the behavior of nation-states anywhere on the planet if they transgress the rights of persons or if smaller states are unable to defend themselves against the reprehensible behavior of other nations.
- *A Planetary Legislature* needs to be elected by the people of the world, not by nations. It is very important that a world constitutional convention be established to create such a body. Most nation-states have such a lawmaking institution, but not the planetary community itself. This may seem utopian, but there is a critical need for it.
- *A Planetary Environmental Monitoring Agency*. This would censure and fine those nations that are egregious in their actions that do severe damage to the environment. The conscience of humanity also needs to protect the atmosphere as well as the rivers and streams of the world. The Kyoto Treaty on climate change of 1997 was an international effort to ban excessive carbon emissions, but it did not go far enough. Similarly the Copenhagen conference of 2009 failed to gain acceptance of a worldwide treaty to enforce compliance.
- *A New Planetary Bank*, which would (a) provide economic loans and assistance to developing countries in the third world with far greater resources than the existing World Bank has, and (b) regulate financial exchanges through principles of sound banking.
- *A Planetary Regulatory Agency of Transnational Corporations*. This would protect the world community from unregulated international corporations, which are often predatory in their practices. It would regulate

practices that escape the control of any nation-state. The G8 or G21 meetings of the major economic powers already provide some guidelines. Such a new agency would encourage free trade and seek to reduce tariffs imposed by some states to protect their industries—no doubt a difficult problem to solve.

- *A Planetary Multinational Peacekeeping Force* is essential. This would oversee security and sanction misbehavior by states. This already in fact partially exists within the United Nations. Unfortunately, its use can be vetoed in the Security Council by any one of the major powers. Thus the role of peacekeeping forces needs to be strengthened in the future if they are to be effective.

In all of this, the planetary community needs to carefully guard against the emergence of an all-powerful world government, one that would undermine the rights of individuals and nation-states. Hence, the powers of a planetary government *should be clearly limited by a new system of checks and balances.*

- *A Planetary Parliament elected by the people of the world* would legislate measures deemed essential by a majority of the elected representatives. This is in addition to the United Nations Assembly, which represents the nations of the world and is unable to function as a legislative body, since it represents independent states, not the people of the world. A planetary constitutional convention needs to be convened to draft a constitution for the world community. The planetary parliament would supplement the United Nations General Assembly as a second house, similar to the United States Congress, which has a Senate in which two senators are selected from each state irrespective of population, and a House of Representatives (Parliament), which is elected on the basis of population.
- *The Planetary Parliament* would elect an executive cabinet to implement its decisions.
- *The Planetary Authority will need powers to tax*—say, 1 percent of the gross national product (at first) of each nation or region.
- It will be concerned with encouraging technical assistance and capital investments to assist emerging nations.

- It will encourage universal healthcare.
- Similarly for universal education.
- Another important prerequisite is the need to encourage and facilitate the continued interaction of the peoples of the world by trade and commerce, emigration and immigration, scientific and educational exchange programs.
- This entails *open access to the media of communication* and the right of access to cultural and educational programs.

At the present time, many nation-states impose restrictions and censorship. Many fundamentalist regions, whether religious or ideological, need to allow the right of dissent, the right to read and communicate, freedom of speech and expression. Fortunately, the emergence of the electronic media, especially the Internet, facilitates the ability to communicate a wide range of ideas and values, and this can add to the sense that we are all part of a new planetary civilization. We are each *citoyens du monde* (citizens of the world). These transnational institutions can contribute to human bonding on the global scale.

The above are some of the reforms recommended in detail in *Humanist Manifesto 2000*. I submit that they need to be adopted by the people of the world, but this is possible only if *planetary ethics* is widely understood and accepted.

The human species has reached the point where the emergence of transnational institutions can best develop a sense of loyalty to the planetary community. It will undoubtedly be a gradual process, but what better way to bring into being important reforms in the current system of nations than by developing a new planetary identity as essential to fulfilling the common good. Loyalty to the family of humanity on the planet Earth would in time transcend separate loyalties to nation-states. One-worldism would supplant nationalism, which has been so destructive of shared human values, especially in times of war when fiendish forces have been unleashed.

I can appreciate the fear that we may be creating a new totalitarian form of government that will repress liberty. I am surely cognizant of the need for a new constitution, specifying checks and balances, allowing maximum decentralization of governance on the national, regional, and local level and guaranteeing the rights and freedoms of individuals and the diverse cultures of the world. The optimist says that it is possible to bring into being a new planetary authority, but the realist is surely aware of the possible dangers, as well as the positive advantages.

COUNTERPOINT

I have in the above discussion confined myself to a short- and mid-range timescale. Who can predict with any degree of certainty the controversies and conflicts that will appear in the distant future? Some possible scenarios easily come to mind. I will list them in no particular order or priority.

First, *the violent conflicts and wars between nation-states are bound to continue* until a new transnational system of law is adopted by the planetary community. Nation-states are armed to the teeth, and a great part of national budgets is devoted to "defense." Many of these powers possess weapons of mass destruction. The problem humanity faces is the concept of "national sovereignty," and until that provincial idea is limited by a new transnational order, and a peacekeeping force becomes effective, any adventurous government or power-hungry leader may be tempted to invade neighboring states and engage in domination "in the national interest." Historically, some states possessed sufficient military force so that they could become the predominant power in parts of the world—England and France in the eighteenth and nineteenth centuries in their colonial empires, and the United States in the twentieth century. Germany and Japan (and their allies) contested this hegemony, and their actions precipitated two world wars.

Russia's challenge to the West in the post–World War II period failed. New challenges from nations such as China may very well confront the world community. The scourge of wars is thus likely to continue to beset humankind in the future, especially given the possible discovery of new weapons that may suddenly give a nation new temptations for military adventures—unless new transnational institutions are created to adjudicate differences.

Second, *economic competition* between individuals, corporations, states, and regions is ongoing. There is the continuing quest for new markets and natural resources. One can imagine that the decline in natural gas and oil supplies will impel economic interests to seek new areas for exploitation, which may lead to wars of conquest—as European colonial exploitation and plunder of Africa, the Middle East, North and South America did in the past. Today the recruiting of cheaper foreign labor is another impulse for imperialistic ambitions. New living space for large countries may tempt them to seize other lands. These problems need to be mitigated in the future.

Third, *racial and ethnic xenophobic antagonisms* have always inflamed human passions against foreigners. Human beings since the earliest times have feared those who were different from them. Witness the oppression of native populations in the Americas, as Indian tribes were expelled from their lands, starved, or slaughtered. Racial, ethnic, or tribal hatred has led to bitter genocides as in the Nazi extermination of the Jews and Gypsies in the infamous Holocaust of the 1940s, the mass killings in Rwanda, and the ethnic "cleansing" that raged after the breakup of the former Yugoslavia. The seizing of blacks in Africa, their transportation to the Americas in slave ships, and their forcible enslavement on Southern plantations led to a bloody civil war. Today fear of Asians is widespread; similarly for hatred of white men as colonial masters. One solution to racial hatred is miscegenation, though the children from such unions are themselves discriminated against. We can only hope that racial distinctions will in time disappear in a new global melting pot.

Fourth, *religious warfare has embroiled humanity* for millennia. Those who did not accept the prevailing religious faith and practices were tortured, exiled, or killed. The Koran and the Bible have been used as instruments of oppression; inquisitions, holy wars, and discrimination have been pervasive in the name of God. This has not only led to fanatic religious wars between Christians and Jews; Muslims and Christians; Muslims against Jews, Hindus, Buddhists, and others, but also between Roman Catholics and Protestants as well as Sunnis and Shiites. Fundamentalist creeds have been instilled in children from their earliest years: they are baptized, circumcised, or confirmed at puberty. All of the major institutions of society—the educational, legal, and economic structures—were conditioned by religious bias.

Freethinkers have criticized such indoctrination and have refuted the God narrative. Will religious wars continue in the future, or can humanist ideals and values ultimately prevail? That is the great question for our time and in the future.

Fifth, *disparities in income and wealth* between classes will undoubtedly continue to inflame societies. This is an internal conflict that all societies and states have suffered. The revolt of serfs, workers, the dispossessed, and their suppression by landowners, plutocrats, industrialists, or financiers is a familiar litany in history. In the nineteenth century it led to strikes and walkouts and to the development of labor unions. In democratic societies such disparities provoked efforts at reform. Marxism led to workers' revolutions in Russia, China, and elsewhere. The Marxist revolt was international in scope, placing in opposition the wealthy Northern industrialized and technologically advanced societies and the developing third world. This involved ideological battles about unfair distribution of income and wealth. It eventually led to an emphasis on growth rather than on redistribution and the introduction of science and technology to improve productivity and increase the wealth of poorer nations, so there is more wealth available for everyone. Some disparities, at least in affluent economies, have been reduced by taxation and the implementation of welfare policies, though there are still billions of poor people on the planet—from Haiti to Bangladesh—who barely subsist, and this includes poor nutrition and a lack of healthcare. These disparities will most likely continue to exacerbate conflicts.

Sixth, *sexual oppression of women persists* in many areas of the world. This discrimination is embedded in many patriarchal religions. Many affluent democratic societies have afforded equality to women—but they still have a long way to go until it is fully implemented. The suffragette movement in the West had extended the right to vote to women. But in large parts of the world this is not the case, and the women's liberation movement has an ongoing agenda to work for full equality between the sexes. In some countries, women are considered lesser persons than men; they are controlled by their husbands, fathers, or brothers and are forced to submit to their will. This issue will undoubtedly fester in the future and provoke struggles for gender equality.

CONTRA-COUNTERPOINT

In the previous section I recounted six areas that have inflamed conflicts in the past and will undoubtedly do so in the future: (1) wars between nations, (2) economic competition, (3) racial and ethnic antagonisms, (4) religious disputes, (5) disparities in income and wealth, and (6) the sexual oppression of women. Some maintain that these perennial disputes occur because they have their roots within human nature. According to this theory, wars and competition are stimulated by the "instinct for aggression." Austrian ethologist Konrad Lorenz said that the instinct is inherited and endogenous. Lorenz studied greylag geese and jackdaws for evidence of this instinct, and he generalized the same instinct for humans. Imprinting was an additional factor in transmitting such behavior to the young.[3] But species imprinting is different from human acculturation or parents teaching their children to hate this or that group, a continuing cause of conflicts.

Nevertheless, it is undoubtedly the case that many impulses and tendencies in the human breast give rise to many such disputes: these would include the quest for power and dominance, greed, hatred, prejudice, envy, jealousy, lust, and other motives. It is important to note, however, that whether these tendencies run riot depends on cultural conditions. Societies have discovered ways to control or redirect such destructive tendencies. We are not the victims of our instincts, but they can often be mastered by reason. The instinct for aggression is manifested especially in the males of a species, though it is expressed by females as well. This genetic disposition is a result of natural selection. It is stimulated by raging testosterone levels, which impel males who are rivals in this pursuit. Aggression also plays a key role in survival strategies and in hunting for game.

Without such an instinct, humans would be reduced to pacifistic grazing mammals. But as hunters who kill, humans are able to marshal their aggressive skills. These raw instincts are tendencies, but they can—up to a point—be modified or controlled by reason. Moreover, it is possible to sublimate or substitute other modes of behavior that are less destructive. Aggression is a contributing factor in many creative endeavors. Humans seek to enter into the world and change it. Pluck is vital to all great enterprises. It is allied with our critical intelligence to cope with challenges and dangers in the natural and sociocultural environment, and it leads to inspiration and invention to find bold new outlets.

For example, competitive sports have been played in virtually every culture to test the *competitive* desire to win, and this is a wholesome venting of aggression that might otherwise lead to hostile violence. We reward our great athletes who demonstrate their skills, and we confer upon them prizes, rewards, and adulation.

The Olympic Games of ancient Greece brought together the best athletes of the various city-states to compete in track-and-field contests of various sorts. They were heralded, not killed, in the process. Thus, even though many of these contests, such as those between gladiators, were very brutal and sometimes led to death, many contact sports today may be dangerous, such as boxing, wrestling, and auto racing, and may result in serious injury or death. Nonetheless, the instinct for aggression had been channeled to constructive aims, good for the warriors and the society. Today we reward great golfers, boxers, tennis players, and chess masters for their physical and mental performances.

The same thing is true on the larger scale for contests between teams of athletes: soccer, football, rugby, hockey. Clubs compete ferociously for a cup or banner or prize; and the top players on professional teams receive huge salaries. But equally fascinating is the impact of such contests on the spectators. I myself love to go to football and hockey games—often physically violent, though the players are well padded in most such sports. The sports fans are often carried away by the passion of the game, and many of them idolize their champions. The cheerleaders on the sidelines egg the fans on, and often there is overwhelming zeal, especially if there is a rough contest or our favorite team is walloping the other side.

Compare this with the actual warfare between nation-states, which can be unbelievably brutal. Invading armies kill everyone in sight, and they engage with abandon in pillage, rape, and brutal slaughter. The enemy is dehumanized and no mercy is shown. This was the case during the Peloponnesian wars between Sparta, Athens, and Persia; and it was true as Genghis Khan sacked Rome, and the Roman armies conquered supine nations—they utterly destroyed Carthage. In World War I, millions of French, British, and American troops fought all-out in the trenches against German and Austrian soldiers. Each of the nations at war aroused patriotic frenzy to support the soldiers or sailors on the front lines. Propaganda whipped up passions against the enemy and for their boys, and millions lost their lives or were maimed. Warfare in the Second World War was so intense that civilians were bombed and offensive actions went beyond the confines of the battlefield to open cities. The Nazis killed innocent populations on the Eastern Front with impunity; Stuka dive-bombers destroyed open cities everywhere; and in the West the Allies retaliated by destroying Dresden; Hamburg; and, in Japan, Hiroshima and Nagasaki.

Thus it is crucial for humankind to create new institutions to protect the world from aggressive nations and find peaceful ways to negotiate and compromise on our differences.

Although the instinct for aggression needs to be redirected to alternative outlets, it is important to distinguish it from the achievement motive. We need to encourage individuals to realize their highest talents to the fullest, to break new ground for themselves and their careers but also for the organizations of which they are a part—to achieve excellence in our department or university, research institute or company, city or club—and to develop team spirit in doing so; and we ought to reward individuals or units that do so. They deserve the accolades of public acclaim, and this can strengthen their resolve to succeed and achieve for themselves and the common good.

Human culture has evolved over the millennia a set of social, democratic, and transnational institutions to protect humans from aggressive behavior and to contribute to the progressive improvement of the human condition. Thus, although naked human apes carry with them the biogenetic predispositions that enable them to survive, cultural institutions and human values have likewise emerged to restrain our native instincts and impulses, and to channel them to constructive ethical norms. These enable us to develop a more peaceful, prosperous, and cooperative existence in which good will and harmony can prevail. It is known as *civilization*.

AN AFFIRMATIVE ATTITUDE: THE *GOOD WILL*

How does this affect the individual person? In my view the best option for humans living in a plural naturalistic world is to adopt a positive attitude toward life, an affirmative life stance of the person of good will. Many pessimists will no doubt be furious at my rejection of their ultimate nihilism, bathed in pathos and gloom. I consider them to be boors!

This approach to life is based on a normative attitude; it is an ethical approach to living. For the secular Neo-Humanist, life is or can be good. This is the primary value for naturalistic humanists. Since we have no illusions about salvation in an afterlife, the only thing that we have is *this life in this world*, and we can strive to realize the highest potentialities of which we are capable. This is priceless. It affirms that *I am a living, breathing, passionate, and rational person. I wish to live as fully as I can*—on my own terms where I can, as an individual—though sharing the goods of life with other sentient beings and expressing a good will toward them.

The good will applies to my own self as well; each person has an obligation to satisfy their own needs and wants, to realize their hopes and aspirations, to achieve a meaningful life of satisfaction, exuberance, and creative joy. It also includes a person's capacity to love other human beings, to express genuine affection toward them, to be loved and appreciated by them.

We should consider life an adventure; I am interested in exploring a wide range of experiences and living with intensity. I also need to be reflective of the things that I wish to do, and I strive to make rational decisions as best I can. Thus, a person should exude a positive attitude toward life and toward oneself, recognizing one's limits and shortcomings and always attempting to improve, but also expressing a good will toward others: one's spouse or companion(s), one's relatives (children, parents, siblings), and the many friends and colleagues a person has met in life, worked with, and played with.

In regard to other human beings whom a person has known, one should feel *good will toward them*. We wish them to do well, to succeed in what they undertake, to thrive and prosper, to be happy. Such a person is truly happy if they are happy, and has conquered jealousy and envy as much as possible.

This expresses the affirmative outlook of the ethical humanist, who appreciates nature yet will use creative thinking to solve problems, enter into the natural world and use it (within reason) for the fulfillment of our obligations to future generations yet unborn, to other species, and to the preservation, as far as we can, of the natural ecology.

Such a person is morally considerate of the needs and interests of others. What a sharp difference with the egocentric, selfish, competitive, grasping person who will stop at nothing to amass power, riches, and status. I concur that we should be achievement oriented, but we need not abandon the basic principles of decency.

The affirmative attitude is optimistic, positive, and courageous, recognizing our limitations but also our great opportunities, and is usually considerate of the interests and needs of others.

WHAT ARE OUR ETHICAL IMPERATIVES?

1. There are many general *prima facie* ethical principles. These include the second *categorical imperative* of Kant, which is based on reason; namely, that we should always consider persons as ends in themselves and never as merely a means to an end.
2. To this we need to add another vital imperative: the *empathetic imperative*, which is rooted in feeling and emotion and which is motivational. It is the union of ethical reason *and passionate empathy* that is a vital source of our feelings of moral responsibilities and duties toward others.

What is the justification for *empathy*? It is intrinsic to who and what we are as humans. It defines us as moral beings, potentially capable of moral choices. This is rooted fundamentally in our moral intuitions. Human beings, as a product of evolution, possess potential moral propensities. Whether these will be realized depends on the cultural environment. A person's moral sense comes into fruition when nurtured by the community. There has been a progressive evolution of moral awareness. Empathy can come to fruition only if it is *internalized* and supported by love and affection, which the infant, young child, adolescent, and adult is able to receive and in time to reciprocate. It is not enough to love oneself, to develop the norms of self-respect, but to genuinely have an appreciation and affection for other humans, no matter what their gender, age, ethnicity, race, creed, or nationality. The basic civic virtue of democracy is that *each person* should be considered as *equal in dignity and worth*. This cognitive rule is based on reason. To be effective, it needs to be supplemented by *the empathetic imperative*—humans need to genuinely feel an affectionate bond with other sentient beings. This is deeply grounded in our feelings ("the heart"), passionate in source and intensity, but it is also justified rationally.

The basic point is that we need to develop and instill in young children affirmative traits of character; and this should be twofold: Its first source is the affection and love of parents and teachers, kin and friends, the ability to love others and be loved by them. Out of this grows the motivation to act morally. The second source is the gradual development of the capacity for moral reasoning, recognizing its importance in solving moral dilemmas and in framing moral judgments. A person needs to compare and appraise alternative courses of action in situations. But there is a difference between the cold, calculated reasoning process in solving technical or tactical problems—the instrumental rationality that is so essential in making choices—and our attitudes that are infused with internalized feelings of empathy. Intellectual agility needs to be supplemented by *caring*, and it is this psychological disposition that transforms an instrumental rational decision into a moral choice, where the unity of reason and passion makes for a rational-*passional* choice, in which a person does what he or she considers to be right for reasons that are carefully evaluated but that are also infused and inspired by empathetic feeling. It is the union of reason and compassion that defines a choice as truly moral, bolstered with deep feelings, and capable of motivating a person to so act; that is, the decision spills out into *praxis*, or behavior. It involves not only our cognition but also our blood and guts, feelings of sorrow and tears, delight and enjoyment, desire and love.

The key point is that although the naked human ape has the capacity for moral behavior, whether he or she actualizes the moral potentialities depends on the culture in which that person lives and fulfills who and what he or she will become. Morality is rooted in biology but comes to fruition in culture. Although there is a body of moral imperatives intrinsic to us—at least potentially—how these come to full bloom and are modified and supplemented depends upon social institutions and the cultural milieu. And there are always new principles that evolve and need to be applied. The normative principles and values of ethical culture have evolved over time, and new principles have been discovered and/or are stretched to apply to new sociocultural contexts. Thus culture is coequal to biology in providing grounds for moral behavior. Although moral behavior has biogenetic roots, its consummatory realization bears fruit in a cultural context.

THE EMPATHETIC PERSON

The attitude that an empathetic person of good will expresses toward others is cordiality, comity, amity, cheerfulness, and generosity. Such a person is sympathetic to their problems if they have any. He or she tries to be helpful: kind, compassionate, and considerate. The empathetic person wishes to console them if they are suffering, to assist them if they need support, and to assure them if they need encouragement. Out of this grows the charitable acts that humans perform to help others in need or distress.

The attitude here is genuine caring for those who need help, being beneficent, if one can, toward them and benevolent in attitude and moral concern. A sympathetic person feels very deeply that if they prosper, then he or she prospers. Such a person wishes to cultivate an affectionate regard for children who are bubbling over and can best be reassured if he or she appreciates their youthful exuberance; but it also applies to elderly people who are grateful for a kind word, a witty remark, a thank you for what they have done.

The same attitude needs to be expressed time and again to the persons one lives with on an intimate basis, to hug and adore them, to tell them how much one needs and loves them. It also is expressed to neighbors in the community, who may appreciate a friendly smile or a nod of recognition. It applies to the women we meet every day—a compliment on the way they dress or look—polite manners toward them are well received. Similarly for the store clerk at the cash register who is tired and works hard. If the clerk's name is known, thank her personally for her assistance and service.

We should also be pleasant to the men we meet and share their concerns. If they deliver a package, salute them with a gesture of appreciation. Men should respect other men and become their friends; they should be considered not adversaries but neighbors.

Every person is different, unique, and idiosyncratic. We should try to find the good that they have done and compliment them, or try to find common ground if we can.

A genuine affectionate regard should apply to every individual, whether male or female, young or old, poor or wealthy, no matter what their station in life or their race or ethnicity. An empathetic person should try to find beauty and dignity in the African, the Chinese, the Anglo or Hispanic, the Italian or the Indian. *A person is a person is a person*; we should respect them and have some genuine concern for their interests and well-being.

Young students especially need encouragement as they embark upon their unique paths in life. Make suggestions if you can, bolster their confidence, urge them to persevere in achieving their goals, wish them well, tell them not to be discouraged; encourage them to have some confidence in their own talents; lend them a helping hand if you can. The best salve for the discouraged person is a compliment, a handshake, an embrace, and, above all, a smile. Advise them to realize their goals and dreams and not to be overly frustrated by adversity. There is no substitute for a good will, a well-intentioned gesture, or a kind and sympathetic word.

Good will is genuine if it flows from internalized empathy felt toward others and a sense of altruism, that we should help them if we can, without thought of favor or advantage. Good will is *good in itself*. It is the expression of a qualitative experience, which, if sincere, is good to the giver as well as the recipient. In one sense, to be able to bestow affectionate interest in others is far better than to receive it—it is the true testimony to the realization of humanist morality.

If religions of the past demanded obedience for God's sake, then to do a good deed *for its own sake* is its own reward. It is *intrinsically worthwhile* for those who have the capacity to express it.

COUNTERPOINT: COPING WITH PERSONAL TRAGEDY

Have I left something out of this optimistic defense of the life of good will? Surely life has its tragic dimensions. So many poor souls suffer unbearable hardship, and so many endure lives of deep pain and sorrow.

All human beings are capable of experiencing bountiful times of quiet satisfaction and hedonic pleasure. Yet all too many encounter bad luck and hard times. Some may be confronted by mean-spirited or malevolent people. Others may live in a repressive society in which their deepest desires and fondest hopes remain unfulfilled; even their basic needs may be unsatisfied. Many suffer hunger, poverty, deprivation, illness, and torment. It depends on what sky they were born under and the social system and the time in which they live. If one is a slave in ancient Egypt, a serf in the steppes of Russia, or a black person in the Southern states of America before the Civil War, one must do the masters' bidding. The parameters of a person's existence are defined by the sociocultural environment—whether an aristocrat in eighteenth-century England, an indigent farmer in nineteenth-century Ireland, or an impoverished twenty-first-century American student struggling against all odds to go to college. It is clear that there are times of great deprivation and misery that may limit a person's capacity for good will and creative achievement.

But how we live is also a function of our beliefs and attitudes. One can lead a full life in spite of adversities. In driving through Morocco several years ago, I passed through a small town in which the children were laughing and frolicking in play. Surprisingly, they all had blond hair and blue eyes. How incongruous. I asked our Moroccan guide who they were and why they seemed so gay. They were Visigoths, he replied, a Germanic tribe from northern Europe that had invaded the area many centuries ago, were marooned, and remained there ever since. They are very poor, said the guide, but they know nothing better, and they are happy.

Living in ancient Athens during its Golden Age or in Italy during the Renaissance did not preclude many individuals—even if impoverished—from realizing lives of creative fulfillment. We today possess comparative historical knowledge of earlier epochs. Thus, the horizons of a person's achievements are influenced by the attitudes and outlook he brings to bear.

Some individuals who are aware of earlier historical times, such as Miniver Cheevy, found that this life is a vale of tears.

> Miniver Cheevy, child of scorn, grew lean while he assailed the seasons.
> He wept that he was ever born, and he had reasons.
> Miniver loved the days of old when swords were bright and steeds were
> prancing. . . .
> Miniver Cheevy, born too late, scratched his head and kept on thinking;
> Miniver coughed, and called it fate, and kept on drinking.[4]

Imagination can arouse despair and unhappiness. It can also expand the realm of expectation and possibility, and this may be a source of fantasy and delight. It is clear that wherever you live depends on the period of history in which you find yourself, including your philosophical, religious, and moral beliefs.

There have been times of great tragedies in the past when it was difficult to find safety. Some of these were cataclysmic in ferocity, such as the plagues and pestilences that have engulfed a region of the world—the great plague of Athens in 430–27 BCE; the Black Death (bubonic plague) that swept Europe from 1348–1350; or the influenza epidemic called the Spanish Flu in 1918, which took millions of lives. There have been devastating famines, such as in Ireland from 1845–48, the great fire of London in 1666, the San Francisco earthquake of 1906, and the numerous earthquakes in 2008–10 in Indonesia, Haiti, and Chile. Such destructive events disrupt people's lives and are ever present through recorded time. Likewise, humanity has experienced devastating wars throughout its history that destroyed so many lives and untold property; suddenly one's world was sacrificed to the turbulent events that swept everything in their path.

What are the effects of these unbearable tragedies on the personal lives of the individuals who suffer them? Everyone experiences disappointments and defeats at various times in life. Surely not all of our hopes and dreams can be fulfilled. A person may be afflicted with an intractable disability or incurable illness, or suffer the pangs of unrequited love. He or she may lose a loved one, be dismissed from a job, become bankrupt, fail in school, or be betrayed by people who were counted on.

Not everything turns out well; we do not succeed in all that we strive for. Thus we must learn to live with misfortune and take a balanced view of what life holds in store for us. We need to balance the agony and ecstasy of living, the fear and trembling with the joy and exhilaration, the good times and the insufferable ones. It is when bad times *persist* and conditions become *unbearable* that a person may be said to live a life of tragic dimensions, especially if there are no escape hatches.

A former editor I knew was struck down with Lou Gehrig's disease; she became completely paralyzed. There is the awful case of the soldier who was wounded in war, lost both arms and both legs, and was rendered blind—hell could not be worse. These cases of profound tragedy evoke our pity and empathy.

Aristotle, in his *Poetica*, presented a classical theory of tragedy as revealed in drama and poetry. He defined this as the downfall of a great man who, due to some defect in character, made wrong decisions. These led to his *dénouement*—the unraveling of the plot that ended in his eventual defeat. In this scenario the person is responsible in some sense for the tragic consequences that emerged as his life unfolded.

There are great tragedies that have befallen historical personalities—such as the assassinations of Martin Luther King Jr. by a distant gunman virtually at the moment of King's greatest triumph, or of John F. Kennedy. Here there is an affinity between triumph and tragedy, though their deaths were not due to their failings.

An eloquent illustration of the wedding of triumph and tragedy was the assassination of Julius Caesar in 44 BCE. Caesar had led his army across the Rubicon, embroiling the Roman Republic in a civil war in pursuit of his rival, Pompey, to Egypt, where Pompey was defeated and beheaded. Caesar battled his other rivals in North Africa and returned to Rome in triumph. He sought to consolidate his power and declared himself dictator in perpetuity. This infuriated his colleagues in the Senate, who conspired against him. They were fearful that Caesar had so undermined the Roman Republic that he must be stopped. And so they gathered around and stabbed him, including his friend Brutus. "Et tu, Brute?" cried Caesar. Here a great hero, due to his lust for power, is brought down in tragic defeat. As he lay dying at the base of Pompey's statue in the Roman Senate, others continued to stab him.

A person's life is analogous to a work of art. Much of a person's lifework is a product of chance and contingency. But some of it remains under the control of the person's choices and actions. Like an artist, one gathers the materials; puts them on the canvas; and gives the painting color, tone, line, form, and shape. Every career is a product of personal decisions and the actions undertaken to realize them, and in spite of adverse social conditions, he or she can often fulfill many of his or her dreams. These all add up, and they comprise a person's lifework. There are key choices—such as the decision to marry, to select a profession, or to decide where to live. And there are key turning points, such as a bitter divorce, an irreconcilable controversy, or a battle to the death. These are done within the limits of the context of one's *lebenswelt* (life world). Nonetheless, a person's supreme work of art is her or his own life. The parts are splashed together and interwoven, harmonious and ordered, or cacophonic and disordered, and at the same time they define who a person is, one's own values, desires, beliefs, and attitudes. The greatest of tragedies are due in one sense to a person's failure to realize his or her deepest dreams and desires—due to cowardice, deceit, jealousy, hatred, envy, greed, or errors of poor judgment—or, in so many cases, because of intractable obstacles in the environment or the culture.

So the gnawing question is: *How shall a person cope with tragedy?* Those that are accidental and contingent are *not* always within a person's power to prevent or to change. Those that a person has oneself created provide *some* capacity to change, for which he or she feels responsible. Though we are buffeted by the winds of fortune that surround us, by good luck or bad, we are nonetheless in some sense masters of our fate. *What will be* depends in part on *what we decide to do*—though chance and turbulence always challenge us.

The key question is: *Can a person cope with tragedy?* Those that are contingent we cannot always control. Those in which a person possesses some power to alter the course of events are intrinsic to one's attitude and outlook and are to some extent controllable. However devastating a tragedy that a person has suffered may be, one needs to go on living and to adapt to new circumstances. Thus the cardinal virtue is the *courage to persist*, to *accept what I cannot change*, and to *change what I can*.

Many survivors of the terrible tragedies of the twentieth century—the Holocaust, tsunamis, the nuking of Hiroshima and Nagasaki, the world wars and the all-too-frequent smaller ones—suffered unbearable destruction of their *lebenswelts*. Many survivors of these disasters were so weighed down by them that they were unable to overcome the tragedy. Yet others were able to do so. They learned that *the secret of life is to fully embrace life itself*, to exult in whatever is left, and to go on living in the awareness that *every moment in time is precious*. These are the heroic human values that are testaments to *the will to survive and thrive*. Fight, fight against the dimming of the light! (to paraphrase Dylan Thomas).

The will to live should embody some *stoic resignation* to accept as given that which I cannot ultimately change, but also the *moral resolve* to do whatever I can to improve my life as best I can, and to enjoy and luxuriate with intensity in what is left. Some people cannot overcome the adversities they have suffered, but others are able to rise above and continue their determination to overcome and even achieve new goals. Thus the lust for life supersedes any hesitation. A person's ultimate concern is to affirm *life against death* and to cherish it for its own sake, sharing the intrinsic goods of life with others.

GRAND FINALE

ETHICAL CULTURE: IN RETROSPECT

Where do we end up? If we go back to the *Overture* of this book, the reader will recall that I began writing it while I was sitting in my study overlooking the French village of Mouans-Sartoux. As twilight set in, the skies began to darken, and it became pitch-black outside, aside from the artificial street and house lights. The Moon, stars, and planets also became visible.

It is a year later. I am back in the same location, pondering the dawn. As the Sun rises, I can see the stunning valley below and in the distance the mountains, bathed in the morning light. I pose the existential question about the meaning of existence—as philosophers are wont to do. Is this an occupational hazard? Perhaps. Yet men and women are often temped to raise existential questions about the meaning of their lives in the scheme of things, especially in times of tragedy—death or disease, defeat or despair. Such questions may even be raised at moments of victory and success, while we are enjoying peak experiences. We ask: Is this *real*? Will it last?

I have lived an incredibly active life, full of milk and honey, piss and vinegar, plans and dreams, aspirations and hopes, some of which have been achieved with the outpouring of enthusiasm; some have been splattered on the rocks of failure and betrayal. Thus I raise the question anew in good spirits and with equanimity. I have lived a full life, bountiful and exciting ("Oh, for a dull day now and then!"). What does it all mean, if anything?

Ethical behavior has no discernible roots in the scheme of things, it being independent of human longings for such an anchor. We ask, why be ethical; why do good deeds; why be helpful, kind, considerate, compassionate? This is the plaintive plea of those who have been battered and bruised by the blows of contingent fortune. Indeed, *why* pursue our goals and plans if they are for naught in the long run? Why live? This question may be considered a total non sequitur for most human beings in their prime, especially those who are driven by the lust for life. Is it meaningful for a man like me who is approaching eighty-five years? I can find no ontological foundations for ethics. In a pluralistic open universe, our ethical principles and values grow out of human choices. The ethical realm is relative to human interests and needs as they emerge in different cultural contexts. Nature itself is indifferent to our fondest hopes and dreams, no matter when they occur in historical time.

My response has been suggested at various points in this inquiry: The universe per se has no discernible meaning in and of itself, and surely there is no privileged place in the vast expanding multiverse for me, my sisters and brothers, my compatriots, or for the human species.

If we look back to Act Two, we find the *biosphere* to be overflowing with incredibly rich diversity of life forms, bursting with melodies and rhapsodies, cacophonic squeaks and grunts, exhalations performed, threatening overcast, and the wondrous risings at dawn and settings at dusk of the orange-yellow Sun, a breathtaking night sky lit up with displays of splendor and beauty. I note that we examined the sciences of paleontology and geology, anthropology and archaeology. We discussed the uncovered fossils of extinct species and later the glorious civilizations of the past that are now dead. History demonstrates the abundant evidence of evolvents and emergents that existed in the past and that exist today only as remnants of extinctions. Igor Stravinsky's moving ballet *The Rites of Spring* dramatizes the dissonances and asymmetries encountered in the biosphere.

We view a similar display of turbulence in the physical universe (Act Four), as we peer into the far corners of the cosmos, observing other solar systems, stars clustered in galaxies, and the clustering of galaxies as they merge with others to produce new ones. So there is, as far as we can tell, evolvents and emergents, expirations and extinctions that exist *throughout nature*. These are *generic traits* exhibited in the universe(s) at large. Gustav Holst's moving musical composition *The Planets* displays in sound the fury and harmony of the physical universe.

The ancient view of reality was based largely on the Earth, Moon, and Sun, and eventually the discovery of other planets in our solar system. This is a limited view of nature, which pointed to a fixed order, a perfect realm of space. Yet we now see that this was mistaken, for our Sun and galaxy are rushing through space at unimaginable speeds. It is not a fixed order of harmony or stability.

Nor is there any evidence for the revelations of the prophets of the so-called sacred books who declared that a divine being created this universe for *us*, that in some way our meager human existence is at the very *center* of things, and that we can escape the flux and find an eternal resting place. The preponderance of evidence indicates that each of us is composed of stardust—of atoms, electrons, and neutrons—and that upon our deaths these will be filtered through endless paths of time in a bottomless universe, assuming different forms.

This realization is all the more poignant if we review how *Homo sapiens* evolved (Act Three). It was only through violent competition with other species including *Homo habilis*, *Homo erectus*, and *Australopithecus*, all of whom are most likely descended from our simian apelike forebears. It was sheer luck and perseverance that enabled the human species to survive in a random universe. The story that humankind was created by God is a tale told by credulous men and women in the infancy of the race. This is no longer a viable myth, and indeed someday, in the near or far future, *Homo sapiens* will continue to evolve, taking on new biogenetic characteristics. Alas, *Homo sapiens* may someday disappear, like all other species.

If we look back at Act Seven, a similar tale can be told about the fate of the great civilization that we now live in and for. Will it not, like all of the great civilizations of the past, decline and disappear because of human folly or indecision, or natural mayhem or mishap—following Mesopotamia and Persia, Greece and Rome, the Mayans and the Ottomans, the Habsburgs and the Khazars? These civilizations came into being with high hopes and valiant expectations, and they are now barely noticed in ancient manuscripts, tombstones, or pyramids, stark reminders of the fate of all things: human, nonhuman, and inhuman in the universe. Dmitri Shostakovich's powerful *Leningrad Symphony* was composed in memory of the Nazi invasion and the Russians' heroic defense of the city. It is an atonal rendition of a furious battle waged during the Second World War, and it is a reminder that all things in time will be forgotten in this turbulent universe.

Meanwhile, the myth of God fashioned in the image of Man has largely been discredited—at least for now, though He or She may be resurrected by some future civilization unaware that God is dead because we killed Her, or because it was a false tale. What then is the prospect for humanity, devoid of the old myths and apprised of a realistic understanding of the universe(s) for what it (they) really is (are): an open, pluralistic, forever-changing universe with evolvents and extinctions, mergings and purgings, disappearances and reappearances, resurgents and emergents?

What should be the response of the skeptical scientific rationalist, secular humanist, atheist, or agnostic to this picture of the cosmic scene? What is the human prospect? That is the question that I now wish to explore.

THE HUMAN PROSPECTUS

The Human Prospect always looks forward to the future. However, it is viewed differently in every age throughout the history of culture—from ancient Africa and Asia, Greece and Rome, Europe and the Middle East, North and South America—yet the barbarians have always threatened to overwhelm this human adventure. Sometimes the future looks auspicious and hopeful; at other times bleak. A nation that emerges from a victorious war is soaring with expectation; the one that has been defeated is demoralized. Today, post-postmodern humans live in a rapidly changing global community. Applying scientific inquiry, we have discovered a different universe and different human prospects—unlike any of the contrived religious tales concocted by past generations. It is an open, pluralistic, unfinished, and evolving universe.

There have always been problems in the past, many insuperable. The first existential confrontation that people face is that every human being will someday die, and there is no escape from that—though theistic religions have sought to deny death.

The second awareness new to human knowledge is that the human species at some future point will either become extinct or evolve into something different. This is an eventuality unique for our age to ponder. It is a consequence of the Darwinian revolution.

Another quandary based on our understanding of history is the realization that our civilization will also eventually disappear, be underwhelmed by failure to act or overwhelmed by forces beyond our control. Our civilization understands that it may well be conquered or merged or radically altered; and that which we have created will be buried by the sands of erosion and virtually forgotten.

The final realization—new to human consciousness—is that our physical universe (the planet Earth, our Sun, and the Milky Way Galaxy) will at some point in the far-distant future either explode or implode, fizzle out or become extinct, replaced by other bodies in a vast universe of unending change.

Well, there are at least four possible options that may be taken in response to these bleak forecasts:

- The first option is to *suppress, repress, or forbid these truths to be told.* Have the priests declare them blasphemous, and the despots will censor those truths that are too bitter to contemplate. That has been the response in the past of theologians allied with despots who have endeavored to keep the truth hidden—but this recourse never fully succeeds. That is why freedom of inquiry and free thought will eventually break down barriers to knowledge.
- The second option is *to give up in despair*, adopt the attitude of the *pessimist or nihilist*, declare that the universe has no purpose; civilization is a fraud, my own existence meaningless because death is final. This option abandons any effort to ameliorate the human condition or work for the progressive improvement of society, and it drowns us in sorrow and gloom. The last resort is to commit *hara-kiri*, to which someone might respond: "Here is the door by which you may exit any time. If you cannot find life worth living, it's your fault and your recourse is suicide!"

Now, the above two options are extreme postures; one need not follow them, for there are at least two additional options as possible responses.

- The third option is to adopt the stance of the *complete optimist*, in answer to the first problem (that each individual will someday die). This is the response that humankind has the power to extend life indefinitely, that we can solve the problem of death for all time. For example, there are persons who are so confident in the ability of future scientists to cure the problem of death that they will sign a contract with individuals who have a fatal disease to freeze them until such time in the future as a cure has been discovered, then unfreeze them and get to work to cure them. This is known as cryonics,[1] a Rip Van Winkle theory that an individual can be restored in some future century and come back to live a full life—let us say in the twenty-fifth or thirtieth century. One hopes there will be no electrical outages that will thaw them prematurely (the stink would be unbearable) and that people in some future century will not still speak with a Texas drawl, in Mandarin Chinese, or outdated French.

- The fourth option is that of the *realistic optimist*, and that is to do the best that we can to extend and enhance life. This is the most sensible approach to adopt. Some have even suggested that each person have several clones of his or her body kept in cold storage, and that whenever an organ needs replacement, he or she can take it off the shelf and repair the body, much the same as an auto supply store has plenty of spare parts. This sounds like science fiction that the skeptic would label as pure fantasy, yet who could have predicted the implantation of artificial hearts? Well, none of this is totally beyond the range of possibility; it is difficult to say what can or cannot be accomplished by future scientific technology or the advanced state of medical science. We most likely *will* be able eventually to extend life and improve its quality. This is already happening worldwide, though maintaining large numbers of centenarians-plus would be very costly to society.

The prospect that the human species at some point in the future may become extinct is always a real possibility, given the danger of the proliferation of weapons of mass destruction, such as poison gas in World War I or the explosion of atomic or hydrogen bombs in an all-out nuclear war. This is why a treaty banning such weapons is critical for the future of humankind, indeed for all life on the planet.

The other possible scenario is whether the human species will in time change biogenetically. This is likely to happen, and humans will have to face those changes when they occur. The science fiction film *Avatar* depicts a handsome new quasi-human species. Humans now possess the power to alter the course of evolution by genetic engineering; that is, eliminating unfavorable characteristics, such as genetic diseases (diabetes or Alzheimer's) and enhancing qualities that we consider favorable (such as intelligence and creativity). Of course we need to apply caution in creating designer babies and only develop those talents that we consider morally worthwhile. So there are realistic options to prevent the extinction of the species or its deleterious modification.

As we have seen, a real danger to the future of life on this planet is environmental degradation, such as lethal global warming and pollution. Every effort can and should be made by humankind to protect the atmosphere from excessive carbon emissions and the oceans and rivers from acidification. Still another danger to life on our planet is a possible future impact by an asteroid. Some astronomers believe that we should begin planning space stations that are ready and able to blast any approaching asteroid, deflecting its path. Whether this is science fiction or a real possibility I will leave for the reader to judge.

A gnawing problem that we cannot avoid is the fear that our civilization will someday decline. This is warranted, for social change is ongoing and no one can establish a perfect society that prohibits change, though utopian reformers from Plato to Marx and B. F. Skinner have attempted to do so. My own realistic appraisal is that a new planetary civilization is already emerging, and this will encompass the whole of humankind. This is a task that I believe urgently needs to be realized. This presupposes as a first step the development of the principles of planetary ethics, the first premise of which is *that every person on the planet, no matter what his or her nationality, ethnicity, race, gender, or creed, should be considered equal as a person of moral worth and value.*

Another problem that we cannot escape is *fear* that someday our Earth, Sun, and Milky Way Galaxy will implode or explode. This is a likely possibility. Some optimists wish us to escape to other planets orbiting other stars. This is in the realm of science fiction, and I have discussed this option in the *Intermezzo*. I will not speculate about the future ingenuity of the human species, though as I have stated earlier, a person should not lose sleep over this future calamity. We need to live life fully, here and now, and in the proximate future that we have some realistic possibility of realizing.

As I have already pointed out, the primary function of creative intelligence is that it is an *adaptive coping mechanism*. In this sense it is instrumental in function, though thought may be contemplative and enjoyed for its own sake. This is much the same as the fact that although reproductive organs serve the function of propagating the species, sexuality provides intense pleasure and can be enjoyed *intrinsically* and for its own sake. Similarly, the auditory sense of hearing enables the species that have it to respond to danger, but it also can be used to enjoy the emergent qualities of musical melodies and lullabies for their own sake.

We recognize that although rationality is an instrument of adaptation, humans can develop systems of mathematics, generate scientific theories, and pose philosophical questions that have no apparent immediate adaptive function but that provide in themselves a sense of elegant cognitive appreciation. Human intelligence enables us to develop a cosmic perspective, as I have outlined in this book. Theoretically, science can be pursued to satisfy intrinsic intellectual interests, and it has been developed, satisfying the standards of inductive and deductive logic. Of course, the growth of knowledge has positive instrumental functions that are pragmatically useful for human survival; though they also afford satisfactions that may be enjoyed intrinsically in themselves.

There is a deeper question that I am raising: What are the implications of a cosmic scene without ultimate purpose or order, yet one that is amenable to biological systems that can adapt and adjust to the hazards and uncertainties encountered? In particular, what are the implications for humans engaged in ethical choice?

Clearly we need to develop a set of sociocultural guidelines by which to live, and this is precisely what has happened in the history of humankind. Coevolution provides us with a biogenetic framework, but we have the power to create new cultural arrangements to solve the problems that arise when people cohabitate in groups and are always receptive to change. I suggest that the great challenge is for humankind to abandon the dogmatic/reactionary moralities of the past and create authentic new ethical cultural systems appropriate to the planetary community.

THE MEANING OF HUMAN EXISTENCE: *FREUDE!*

To return to the question "What is the meaning of *existence*?" I respond that it has no meaning in itself; meaning is present only if there is consciousness and awareness. Existence takes on significance for living beings who strive to question their place in the universe and try to cope with and understand it. Our meanings are discovered in the process of living, and they vary from person to person, whether one is a poet or teacher, artist or scientist, explorer or builder, diplomat or healer. Its content is cultural in meaning and form and will vary with the culture, yet it may have a powerful *personal meaning* as well because it helps us to provide a cosmic framework in which to live.

The question "Why should there be something rather than nothing?" is a deceptive question, because the universe is full of matter and energy, individual substances and things, and it is enriched for us by our love for each other and by the passionate nuances of the arts and sciences. These cultural experiences have an emergent reality. It is far more difficult to imagine an empty universe full of nothing than a universe chock-full of entities and systems of various colors and hues, sounds and qualities, relationships and fields, waves and particles, objects and things. "What caused the universe?" is at this stage of human knowledge a question without identifiable meaning because causes refer to concrete objects and fields and may not make sense if applied to everything *überhaupt*. The multiverse is a pluralistic scene or scenes, truly overflowing with diversity and richness, emergents and extinctions. To leap from *the* universe to an ultimate cause is an illegitimate move seized upon in the infancy of the race. Far better at this stage of human knowledge is to assume the posture of the *agnostic* and *skeptic*, confessing our inability to understand that question, let alone respond to it. Nature *is the given* without the need to ground it in something outside itself.

Among the generic traits of nature are its pluralistic qualities: the existence of matter and energy, fields and systems, historicity and individuation, order and disorder, regularity and dissonance, chance and adaptation, turbulence and contingency; and it is a universe where life has emerged and continues to evolve.

Thus the live question that we need to address is whether life is or can be worth living—not only surviving but thriving by realizing our fullest potentialities as persons. This, we are admonished, needs to be examined in the light of our awareness of death. The besetting problem with theism is that it is unable to confront and accept the brute facticity of death, my own death; those of my parents, children, lovers, and friends; and the eventual death of my civilization, or species. For theistic religion proclaimed, in the books of Abraham at least, a grand system of denials—that Moses, Jesus, or Mohammed will rescue true believers, at least from dying, by providing salvation. This is the great deception that has been perpetrated on countless generations of our forebears by the priestly conspiracy to deny death. They are unable to accept the reality of death, and as such, they are congenitally unable to fully realize life. Thus the question that confronts us finally is that death is the equalizer of everyone—the poor person and the wealthy, the baron and the peon, the pharaoh and the slave, the Nobel laureate and the ordinary worker, the child and the adult.

Thus we are brought to confront our own existential plight: What are the implications of accepting death, finally? The fact is that my life and yours—and my civilization and yours—are transient, turbulent, evanescent events. And the only response that we can give is to *affirm* life as intrinsically worthwhile in itself. Each one of us is here as the result of surviving the billions of sperm competing with others to reach the few eggs, penetrate one, and proceed on the hazardous journey to adapt and survive. The astounding feat is that we made it; one of trillions of wasted sperm has fertilized this egg, which eventually becomes me or you.

The ultimate certainty is that each of us will someday die and become extinct. We have a choice—to break down in tears at the transient character of *all* existence, including my own, and to try to seek consolation in the refuge of transcendence—or to accept that fact and resolve with fortitude and courage to become what we want and to work with others to create a safer, more productive, peaceful, and harmonious society in which the *exultant* response is the *lust for life*! Why not the joy of living as the highest option of the human person, who accepts the challenge of living fully as best he or she can, finding some satisfaction, equanimity, and happiness, even exuberance, living with intensity, savoring every quality of life—intellectual discovery, moral aesthetic and social values, romantic love, love of friends and colleagues for each other, as we embark upon the adventure of living! The final choral movement of Beethoven's "Ninth Symphony"[2]—heralding *Freude*—is the grand human finale to a life well lived, reaching jubilation in peak experiences as a testament to the fullest realization of the exuberant life.

This affirmative attitude of good will is its own reward, not in heaven but in human terms, the quickening of our rational powers, our capacity to know and understand, to learn and discover, and the capacity to experience deeply and widely all of the multifarious goods, joys, pleasures that culture and civilization offer, the satisfaction of loving others and empathizing with their needs, helping them where we can, and realizing our own creative capacities with excellence. This, I submit, is the response of the free person, liberated from fixations spun out in the infancy of the race: the world of fantasy that denies the world of nature as we find it.

Humans use their creative imaginations to dream and conceive of new tomorrows; they discover, invent, and create new worlds, and these enable them to find life engaging, interesting, exciting, and satisfying, in spite of the fact that we are part of the great flux of nature. While we are here—however briefly—we can make our distinctive mark upon the world, and this itself is a wonder to behold. Each person has that capacity; to deny it or flee from it, to escape from reason and freedom is folly. The ultimate good is to live every moment that we can and untap while we can the bountiful joys of creativity and fulfillment. And this is intrinsically worthwhile in and for itself. I do not deny the pathos of tragedy, despair, failure, and defeat that so many suffer, yet in spite of that, life can be on the whole joyful. I am not unaware of the exaggerated sense of Voltaire's *Candide* that this is "the best possible world." Yet life is what you make of it, and it is up to each person to find significance in his or her life, and to hopefully find it worthwhile and even joyful.

REVERENCE FOR NATURE

Finally, we need to express a new kind of *natural reverence* for the spectacle of the vault of nature and cultivate a kind of *natural cosmic piety*, stimulated by awe at the magnificence of nature, as the cradle in which we are nourished and from which we can find it meaningful. This mood has been virtually lost to modern men and women and needs to be recovered to put us in some sort of wider perspective: we are part of a boundless sea of astronomical bodies, only a small part of the limitless galactic display.

Primitive men and women, from the earliest days of human infancy, their childhood and adolescence to adulthood and senescence, have been exposed every evening to the naked sky whenever it was a clear night. Humans have viewed the Moon and planets, stars and nebulae, and clusters of galaxies in all of their breathtaking intensity. They were absorbed within a vast canopy that enveloped the Moon, "the Queen of the Night," its "lantern" on the world, casting eerie shadows that we might see changing from a full to half or quarter Moon, or a sliver of white light flitting in and out of the shadows and clouds.

They marveled at the sprinkled stars and nebulae, some as bright as sparkling jewels, others barely visible to the naked eye. Shakespeare described it as "the majestical roof fretted with golden fire,"[3] and Omar Khayyám depicted the arch of heaven, "our inverted bowl they call the sky."[4] They are "the infinite meadows of heaven," said Longfellow.[5] It contains fireballs, meteor showers, and comet tails far away on the horizon and outer space. This is contrasted by winter gales, summer torrents, cyclones and tornados on the inner space of Earth. In the morning the Sun arises, spreading warmth and glow, lights up the sky; but at night or dusk the celestial sphere displays the pockmarked Man in the Moon, bright Venus, reddish Mars, Jupiter and Saturn, the North Star, and the Big Dipper. The Sun warmed their bones and quenched their thirst, by the gentle or torrential rains that fell from the sky, overjoyed at times by the brilliant rainbows and hues that are lit up by the Sun's glow. Humans move from bright day to dark night, and when the clouds covered the heavens above, they sometimes huddled in hushed tones in pitch darkness.

Modern men and women, living in cities and traversing highways flooded with bright lights, very rarely have an opportunity to peer into the abode of the heavenly bodies; they are shielded from the star-studded celestial sphere; and so they go about the business of life, concentrating on mundane desires and narrow purposes, forgetting to look upward at the overhanging cosmic display and to put things in proper perspective. If they do, they cannot help but be overcome by the vast cosmic scene and the insignificance of their own paltry interests: What does it portend? What does it mean, if anything? How does it intermesh with our lives—humans have always pondered these inscrutable questions.

Modern humans, using the instruments of science and reason, can peer far beyond the outer edge with polished mirrors that make up our telescopes and transport us out there. What does this portend *for us*? they ask. Modern astronomy has, since the age of Newton and Galileo, focused on our solar system, and it was inspired by the lawlike regularity of its orbits. But now, following Hubble's lead, we are catapulted beyond our solar system or galaxy to torrents of galaxy clusters. What does this portend for men and women but a new stimulus for a deeper sense of reverence for nature and life?

We are not talking of the ancient anthropomorphic figments of human imagination, the gods and goddesses that were postulated to soothe the aching heart and keep alive a sense of hope about possible promises of a new salvation. Alas or Hurrah—the gods are now dead, have no meaning or truth, heralded by false prophets. The universe is what it is, and there is no evidence of a supernatural realm made especially for us. That illusion is finally shattered by the skeptical eye: these ancient gods are mere figments of our imagination. Yet there still is deep promise within the human adventure, if we would only unleash our response to the challenges, and open up new potentialities for the good life.

Thus whatever we do, we must never forget to look up and stargaze, and this can but only arouse a new sense of awe about nature and a kind of profound reverence for the life that is born of it. The best response we can give is not supplication but buoyancy: to enjoy every moment that we can, to exult and extol the natural world, and to live as fully as we can, realizing our highest talents for creativity and fulfillment.

The theme song of the human species is *joy and exuberance*. And that is our testament to the intrinsic value of life—in itself and for its own sake—in this turbulent unfinished universe, in which we can each find a place, if we have the courage to become what we will, by using our rational powers and expressing the intensity of living. In a eupraxsophy of hope—we can bring practical wisdom to experience the fullness of the bountiful life that we can share with others. And yet, there should always remain for all men and women, young and old, a deep sense of awe at the magnificence of the natural scene in all of its glory and majesty, of which we are, each of us, a part.

NOTES

OVERTURE

1. The "new atheists" have published a number of books that have been characterized under this rubric. Richard Dawkins, *The God Delusion* (New York: Houghton Mifflin Harcourt, 2006); Christopher Hitchens, *God Is Not Great: How Religion Poisons Everything* (New York: Hachette, 2007); Sam Harris, *The End of Faith* (New York: Norton, 2004); Daniel Dennett, *Breaking the Spell: Religion as a Natural Phenomenon* (New York: Penguin: 2006); Victor Stenger, *The New Atheism* (Amherst, NY: Prometheus Books: 2009).

2. Brian Greene, *The Elegant Universe: Superstrings, Hidden Dimensions, and the Quest for the Ultimate Theory* (New York: Vintage Books, 1999).

3. Carlos I. Calle, *The Universe: Order without Design* (Amherst, NY: Prometheus Books, 2009).

4. See Nathan Bupp, ed., *Meaning and Value in a Secular Age: Why Eupraxsophy Matters. The Writings of Paul Kurtz* (Amherst, NY: Prometheus Books, 2012).

ACT ONE

1. See *Science* 324 (June 5, 2009): 1262.

2. See William Whewell, *Philosophy of the Inductive Sciences: Works in the Philosophy of Science 1830–1914* (London: Thoemmes Continuum, 1999); E. O. Wilson, *The Creation: An Appeal to Save Life on Earth* (New York: W. W. Norton, 2007); E. O. Wilson, *Consilience: The Unity of Knowledge* (New York: Vintage Books, 1998).

3. See "Coduction: A Logic of Explanation in the Behavioral and Social Sciences," in Paul Kurtz, *Philosophical Essays in Pragmatic Naturalism* (Amherst, NY: Prometheus Books, 1990).

ACT TWO

1. "Spring" by Gerard Manley Hopkins.

2. Here it is in Old English:

> When that April is with his showers soothe
> The drought of March hath pierced to the roote,
> And bathed every veine in swiche liquor
> Of which virtu engendered is the flowr

3. For two provocative accounts of the universe, see Louise B. Young, *The Unfinished Universe* (New York: Oxford University Press, 1986); and Brian Greene, *The Elegant Universe: Superstrings, Hidden Dimensions, and the Quest for the Ultimate Theory* (New York: Random House, 1999).

4. See, for example, Discovery Earth, "Mass Extinctions," http://dsc.discovery.com/earth/wide-angle-/mass-extinctions-timeline.html (accessed January 31, 2013), and Anthony Hallam and P. B. Wignall, *Mass Extinctions and Their Aftermath* (Oxford: Oxford University Press, 1977).

5. John Noble Wilford, "From Arctic Soil, Fossils of a Goliath That Ruled the Jurassic Seas," *New York Times*, March 17, 2009.

ACT THREE

1. Robinson Jeffers, "The Bloody Sire," in *The Selected Poetry of Robinson Jeffers*, ed. Tim Hunt (Stanford, CA: Stanford University Press, 2001).

2. There is one caveat that I must introduce. It is possible that organic molecules on Earth came from outer space, perhaps existing on a comet or asteroid that crash-landed on this planet, and that showers of organic matter, seeds, or spores led to organic molecules or cells. No doubt a speculative hypothesis, for the predominant view today is that all forms of life on our planet descended from a common ancestor; and the evidence for this is the similarities that these display. But it is not beyond the range of possibility that the life that evolved here, as seen in the present, and the fossil record from the past, may have come from outer space. That is the idea that Francis Crick (codiscoverer of DNA) suggested in his book *Life Itself: Its Origin and Nature*. This is not unlike the panspermia theory, which conjectured that spores came to Earth from some other planetary body or solar system. Many meteorites have been discovered on Earth. There is some evidence for the extraterrestrial origins of life in the Murchison meteorite, which fell in Australia north of Melbourne in 1969. According to scientists who examined the meteorite, it contains amino acids, one of the building blocks of life on Earth. Other scientists have suggested that this meteor may have been contaminated on Earth. The scientists who examined the meteorite said that two types of the meteorite contained carbon 13, and that it could only have been formed in outer space. These scientists suggested that early life on Earth may have absorbed nucleobases and amino acids from meteorite fragments and transmitted the genetic coding to

successive generations. Even if organic compounds in the earliest forms of life showered down from outer space, evolution still is important to explain the descent of different species that may have evolved from them.

3. See Genesis 1–2.

4. Charles Darwin, *The Origin of Species* (Amherst, NY: Prometheus Books, 1991), pp. 49, 59.

5. From George Santayana, "A General Confession," in *The Philosophy of George Santayana*, ed. Paul Arthur Schilpp (Evanston and Chicago: Northwestern University, 1940).

6. Charles Darwin, *The Descent of Man* (Penguin Classics, 1871).

7. There are many fine books on the subjects of life on Earth and the evolution of humans, including Piero and Alberto Angela's books *The Extraordinary Story of Life on Earth* (Amherst, NY: Prometheus Books, 1996) and *The Extraordinary Story of Human Beings* (Amherst, NY: Prometheus Books, 1993).

8. The discovery in 2009 of the fairly well-preserved fossils of a forty-seven-million-year-old primate indicates that it may be the link between humans and apes. Known as *Darwinius masillae*, it is a lemur-like creature; it had fingernails and opposable thumbs; its hind legs suggest evolutionary changes that eventuated in the ability to stand upright.

9. See Jerry A. Coyne, *Why Evolution Is True* (New York: Viking, 2009).

10. Ibid., p. 206.

11. Ibid.

12. Ibid.

13. Ibid.

14. Carl Zimmer, *Evolution: The Triumph of an Idea* (New York: HarperCollins, 2002).

15. Ibid.

16. Ibid.

17. Darwin, *Descent of Man*.

18. See especially Marc Hauser, *Moral Minds: How Nature Designed a Universal Sense of Right and Wrong* (New York: HarperCollins, 2006).

19. Richard Dawkins, *The Selfish Gene* (Oxford: Oxford University Press, 1976, 2d ed., 1989). See also Daniel Dennett, *Consciousness Explained* (Boston: Little Brown, 1991); and Susan Blackmore, *The Meme Machine* (Oxford: Oxford University Press, 1999).

20. Zimmer, *Evolution*.

21. Ibid., p. 275.

22. Ibid., p. 276.

23. Coyne, *Why Evolution Is True*, p. 145.

INTERMEZZO

1. See Roger-Maurice Bonnet and Lodewyk Woltjer, *Surviving 1,000 Centuries: Can We Do It?* (New York: Springer Praxis Books, 2008).

2. Philip J. Klass, *UFOs: The Public Deceived* (Buffalo, NY: Prometheus Books, 1986); Philip J. Klass, *UFO Abductions: A Dangerous Game* (Buffalo, NY: Prometheus Books, 1989).

3. Allan Hendry, *The UFO Handbook: A Guide to Investigating, Evaluating, and Reporting UFO Sightings* (Garden City, NY: Doubleday, 1979).

4. P. Kervella et al., "The Diameters of *a* Centauri A and B—A Comparison of the Asteroseismic and VINCI/VLTI Views," in *Astronomy & Astrophysics* 404, no. 3 (2003): 1087.

ACT FOUR

1. W. V. Quine, *Theories and Things* (Cambridge, MA: Harvard University Press, 1981), p. 21.

2. W. V. Quine, "Reply to Putnam," in *The Philosophy of W. V. Quine*, edited by L. F. Hahn and P. A. Schilpp (LaSalle, IL: Open Court, 1986), pp. 430–31. Both are quoted in "Quine's Naturalism in Question," by David Alexander, in *Philo* 11, no. 1.

3. See David Macarthur, "Quine's Naturalism in Question," *Philo* 11, no. 1:11.

4. Quine, *Theories and Things*, p. 98.

5. Steven Weinberg, *The First Three Minutes: A Modern View of the Origin of the Universe* (New York: Basic Books, 1988).

6. Quoted in Natalie Angier, *The Canon: A Whirligig Tour of the Beautiful Basics of Science* (New York: Houghton Mifflin, 2007), p. 87.

7. Physicist Leonard Tramiel qualifies this statement by asserting that "the uncertainty principle is not a limitation of our technology. Quantum mechanics allows a particle to have many different positions and/or moments at the same time. The uncertainty principle says that the product of those uncertainties has a minimum value."

8. From a talk delivered by Murray Gell-Mann at the XII Center for Inquiry World Congress, 2007. Published in a Chinese-language edition by the China Research Institute for Science Popularization (CRISP), Beijing, China, 2009.

9. Ibid.

10. Ibid.

11. Steven Weinberg, *Dreams of a Final Theory: The Scientist's Search for the Ultimate Law of Nature* (New York: Pantheon Books, 1993).

12. "You and I," words and music by Robert Meredith Willson, recorded by Arthur Miller, 1941.

13. Weinberg, *Dreams of a Final Theory*, pp. 35–36.

14. E. O. Wilson, *The Creation: A Meeting of Science and Religion* (New York: W. W. Norton, 2000), p. 110.ff.

15. Ibid.

16. Philip Plait, *Death from the Skies: These Are the Ways the World Will End* (New York: Viking, 2008).

17. Ibid.

18. Roger-Maurice Bonnet and Lodewyk Woltjer, *Surviving 1,000 Centuries: Can We Do It?* (New York: Springer Praxis Books, 2008), p. 18.

19. The moons of Jupiter are numerous: Métis, Adrastea, Amalthea, Thebe, Io, Europa, Ganymede, Callisto, Themisto, Leda, Himalia, Lysithea, Elara, S/2000 J 11, Carpo, S/2003 J 12, Euporie, S/2003 J 3, S/2003 J 18, Thelxinoe, Euanthe, Helike, Orthosie, Iocaste, S/2003 J 16, Praxidike, Harpalyke, Mneme, Hermippe, Thyone, Ananke, S/2003 J 17, Carme, Aitne, Kale, Taygete, S/2003 J 19, Chaldene, S/2003 J 15, S/2003 J 10, S/2003 J 23, Erinome, Aoede, Kallichore, Kalyke, Callirrhoe, Eurydome, Pasithee, Cyllene, Eukelade, S/2003 J 4, Pasiphaë, Hegemone, Arche, Isonoe, S/2003 J 9, S/2003 J 5, Sinope, Sponde, Autonoe, Kore, Megaclite, S/2003 J 2.

20. A team of four scientists presented the "Nice model": H. Levinson, R. Gomes, K. T. Siganis, and A. Murbidelli, in *Science* (December 3, 2004): 1676. See also "Shifting Orbits Gave Our Solar System a Big Shakeup, Model Suggests," *Science* 325 (July 17, 2009).

21. Lawrence M. Krauss and Robert J. Scherrer, "An Accelerating Universe Wipes Out Traces of Its Own Origins," *Scientific American* (March 2008).

ACT FIVE

1. Stephen J. Gould with Niles Eldredge, "Punctuated Equilibria: An Alternative to Phyletic Gradualism," in *Models in Paleobiology*, ed. Thomas J. M. Shopf (San Francisco: Freeman, Cooper, 1972).

2. Ludwig von Bertalanffy, *Perspectives on General System Theory: Scientific-Philosophical Studies*, ed. Edgar Taschdjian (New York: George Braziller, 1975).

ACT SIX

1. See Carl Zimmer, *Microcosm: E-coli and the New Science of Life* (New York: Pantheon, 2008).

2. C. A. Tripp, *The Intimate World of Abraham Lincoln* (New York: Basic Books, 2006).

3. "Monticello Research Committee Report on Thomas Jefferson and Sally Hemings," Thomas Jefferson Foundation, January 2000.

4. Dmitri Volkogonov, *Stalin: Triumph and Tragedy* (New York: Simon & Schuster, 1991); Dmitri Volkogonov, *Lenin: A New Biography* (New York: Simon & Schuster, 1994); Dmitri Volkogonov, *Trotsky: The Eternal Revolutionary* (New York: Simon & Schuster, 1996).

ACT SEVEN

1. *National Geographic* 216, no. 6, December 2009, http://ngm.nationalgeographic.com/2009/12/hadza/finkel-text.
2. Arnold J. Toynbee, *A Study of History* (Oxford: Oxford University Press, 1939).
3. Will Durant and Ariel Durant, *The Story of Civilization* (New York: Simon & Shuster, 1935–75).
4. Jared Diamond, *Guns, Germs, and Steel: A Short History of Everybody for the Last 13,000 Years* (New York: Vintage, 1998).
5. Ibid.
6. US Constitution, art. 2, sec. 1.
7. Paul Kurtz and Edwin H. Wilson, *Humanist Manifesto II*, 1973.

ACT EIGHT

1. Eupraxsophic pragmatic intelligence refers to conduct where a rational cognitive element has intervened and made us especially sensitive to the real-world *consequences* of our actions. This is wise ethical living.
2. One caveat to this: there are still small tribes in the Brazilian rainforest or in African jungles that have not had extensive contact with the outside world, but these are rapidly disappearing.
3. American philosopher Felix Adler created the first *ethical culture* societies at the end of the nineteenth century in recognition of the need of human communities to develop new ethical societies appropriate to the modern world quite independent of the old-time theistic religions. While there is a powerful insight in his recognition, in my view we need to advance beyond his Kantian ethical scheme to create entirely new transcultural, ethical values and principles.
4. This book was the fifteenth published by Prometheus Books, a company that I founded. Steve was a close friend and colleague.
5. See Paul Kurtz, *Decision and the Condition of Man* (Seattle: University of Washington Press, 1965), Delta paperback (New York: Dell, 1968).
6. Sidney Hook, *The Hero in History* (Boston: Beacon, 1955).

ACT NINE

1. Paul Kurtz, *Humanist Manifesto 2000: A Call for a New Planetary Humanism* (Amherst, NY: Prometheus Books, 2000) has been translated into dozens of languages and endorsed by many international scientists and scholars.

2. The term "World Court" often refers to the International Court of Justice (ICJ) headquartered at The Hague, Netherlands. It was established in 1946 by the UN Charter. Its function is to settle legal suits submitted to it by member states and authorized international organs. Its opinions are only advisory.

A second body is the International Criminal Court (ICC), a permanent tribunal that was established by treaty in Rome, on July 1, 2002, also at The Hague. Its mission is to prosecute individuals for genocide, crimes against humanity, war crimes, and the crime of aggression. It is independent of the UN. By late 2009, 110 states have signed and ratified the treaty (known as the Rome Statute). Another thirty-eight states have signed but not ratified the statute. The United States, Russia, India, and China have not joined the ICC. At present its jurisdiction is severely limited, and only fourteen individuals had been indicted. Efforts to make the ICC universal in jurisdiction were opposed by the Bush administration and defeated. The world desperately needs such a tribunal with broader authority.

3. See Konrad Lorenz, *The Instinct for Aggression* (New York: Harcourt, 1966), and Nikolaas Tinbergen, *The Study of Instinct* (Oxford: Clarendon, 1951).

4. From the poem "Miniver Cheevy" by Edwin Arlington Robinson, 1910.

GRAND FINALE

1. See Rudi Hoffman, "Many Are Cold but Few Are Frozen: Cryonics Today," *FREE INQUIRY* 27, no. 6 (October/November 2007).

2. This is based on Friedrich Schiller's "Ode to Joy," which provided the text for the chorus.

3. William Shakespeare, *Hamlet*.

4. Omar Khayyám, "The Moving Finger Writes," from *Rubáiyát of Omar Khayyám* (n.p., n.d.).

5. Henry Wadsworth Longfellow, *Evangeline: A Tale of Acadie* (n.p., 1847).